The Enlightenment's Animals

The Enlightenment's Animals

Changing Conceptions of Animals in the Long Eighteenth Century

Nathaniel Wolloch

Amsterdam University Press

Cover illustration: Jan Steen, *Children Teaching a Cat to Dance*, also known as *The Dancing Lesson*, 1660-1679, oil on canvas, 26.8 x 23.2 in.
Rijksmuseum, Amsterdam

Cover design: Coördesign, Leiden
Typesetting: Crius Group, Hulshout

ISBN 978 94 6298 762 3
e-ISBN 978 90 4853 932 1 (pdf)
DOI 10.5117/9789462987623
NUR 685

Printed and bound by CPI Group (UK) Ltd, Croydon, CR0 4YY

To Naomi and Jonathan

Contents

List of Illustrations

Preface

This book began as a collection of articles published over many years. These articles, however, have been revised, at times considerably, and combined to form an organic interrelated argument, reflecting the interpretational logic which underlines my view of the history of attitudes toward animals. In revising these initially separate studies it was necessary to omit all of the overlapping material, which has here been mainly consigned to the introduction. If for this reason alone, those who are interested only in one or another of these chapters would do better to go to the original versions. Other revisions have been more substantive, at times reflecting changes in my view of these topics. To these previous studies I have also added a significant new chapter which closes the book, one which puts all the other chapters in a different and more unified light. I have also attempted, as much as possible, not to repeat claims or material from my previous book on the history of attitudes toward animals, *Subjugated Animals: Animals and Anthropocentrism in Early Modern European Culture*. Some discussion of anthropocentrism, as of the Cartesian beast-machine theory and of early modern theriophily, could not be avoided, but I have tried to keep it to a minimum. These topics have received ample attention over the years from myself and many other scholars. Ignoring them is impossible, but neither do they require detailed (re)explanation. The term "theriophily," "love of animals," is used throughout the present book but, as will be explained, in a somewhat broader and looser sense than is usually customary. I have also tried to avoid as much as possible any partisanship regarding the ethical treatment of animals. As a scholar of Enlightenment historiography I do not deny the need for an ethical outlook when writing history. Nevertheless, in today's political climate anything which has to do with the environment, and specifically with animals, tends to arouse passions which are not conducive to a proper historiographical approach. This is more appropriate for philosophers than historians. In *Subjugated Animals* I succumbed to this temptation, to the detriment, so I believe today, of the quality of the discussion. Or have I simply moderated my views with the passing of the years? In any case, the material is presented here to the readers in a generally impartial manner, and they can do with it as they please in terms of how they fit it with their own philosophical beliefs.

The chapters in their original formats were previously published as follows: chapter 1 as "An Interpretation of Early Modern Vivisection," *Zmanim*, 67 (1999), 22-33 (originally in Hebrew; published by the Open

University of Israel); chapter 2 as "Christiaan Huygens's Attitude toward Animals," *Journal of the History of Ideas*, 61 (2000), 415-32 (published by the University of Pennsylvania Press); chapter 3 as "*The Turkish Spy* and Eighteenth-Century British Theriophily," *Eighteenth-Century Thought*, 4 (2009), 67-85 (published by AMS Press); chapter 4 as "Rousseau and the Love of Animals," *Philosophy and Literature*, 32 (2008), 293-302 (published by Johns Hopkins University Press); chapter 5 as "William Smellie and Enlightenment Anti-Anthropocentrism," *Eighteenth-Century Life*, 33 (2009), 45-63 (published by Duke University Press); chapter 6 as "The Status of Animals in Scottish Enlightenment Philosophy," *Journal of Scottish Philosophy*, 4 (2006), 63-82 (published by Edinburgh University Press); chapter 7 as "The Limits of Enlightenment Sensitivity to the Suffering of Animals," in *Knowledge and Pain*, ed. by Esther Cohen, Leona Toker, Manuela Consonni, and Otniel E. Dror (Amsterdam and New York: Rodopi, 2012), 123-44 (originally published by Rodopi, now by Brill); chapter 8 as "Animals in Enlightenment Historiography," *Huntington Library Quarterly*, 75 (2012), 53-68 (published by the University of Pennsylvania Press); chapter 9, in an original version which was more generously illustrated, as "Dead Animals and the Beast-Machine: Seventeenth-Century Netherlandish Paintings of Dead Animals, as Anti-Cartesian Statements," *Art History*, 22 (1999), 705-27 (published by Wiley); and chapter 10 as "Adam Smith's Economic and Ethical Consideration of Animals," *History of the Human Sciences*, 26 (2013), 52-67 (published by Sage). Chapter 11, as already noted, is completely new and does not include any significant previously published material.

The debts incurred over the many years of working on this book, from my days as a student more than twenty years ago to the recent highly professional and generous review and production process at Amsterdam University Press, are too many to even attempt to name in detail. Had I tried to do so, I would have inevitably ended up forgetting some people. I therefore hope that they will all accept this general and collective expression of gratitude.

This book is dedicated to my children.

Introduction

This book outlines a central thesis, which, put simply, asserts that the study of early modern attitudes toward animals, mainly in the long eighteenth century, has unjustifiably concentrated on the history of philosophy and science and has failed to give adequate attention to emerging historiographical and economic conceptions of animals. A concomitant of this traditional approach has been an undue concentration on debates about the physical, and mainly mental, similarities and dissimilarities of humans and animals and also, in many cases, an overstatement of the rise of a modern morally sensitive attitude toward animals in the Enlightenment. In departing from this common historiography, the book begins intentionally with a discussion of the more familiar territory of the intellectual history of attitudes toward animals in science and philosophy but then gradually moves to the history of historiographical and economic conceptions of animals. In this way the importance of the more familiar materials is not denied, but at the same time the novelty of the less familiar materials can be comparatively appreciated. From a methodological vantage point as well, the interdisciplinary nature of the discussion, and specifically the integration of visual artistic sources into the field of intellectual history, is meant to show that the history of early modern attitudes toward animals is far from a limited philosophical or scientific topic.

Not long ago, the study of the history of attitudes toward animals still seemed to require justification.[1] Recent years have seen a growing stream of publications in this field, making any such justification all but redundant. Particular attention has been devoted to early modern attitudes toward animals.[2] This is no accident. It was during the early modern era that the cultural and intellectual consideration of animals gradually assumed its modern form. By the end of the eighteenth century, the way people

1 Delort, *Les animaux ont une histoire*, 7-11; Fudge, "Left-Handed Blow"; Ritvo, "Animal Planet."

2 For a far-from-comprehensive selection limited to book-length studies, see e.g. Thomas, *Man and the Natural World*; Fudge, *Brutal Reasoning*; Fudge, *Perceiving Animals*; Boehrer, *Shakespeare Among the Animals*; Wolloch, *Subjugated Animals*; Fudge, ed., *Renaissance Beasts*; Palmeri, ed., *Humans and Other Animals*; Shannon, *Accommodated Animal*; Muratori and Dohm, eds., *Ethical Perspectives on Animals*; Senior, ed., *Cultural History of Animals*; Robbins, *Elephant Slaves and Pampered Parrots*; Cole, *Imperfect Creatures*; Quinsey, ed., *Animals and Humans*. Also see the special journal issues: Ridley, ed. "Animals in the Eighteenth Century"; Cole, ed., "Animal, All Too Animal"; Meli and Guerrini, eds., "The Representation of Animals in the Early Modern Period." Among older studies see Harwood, *Love for Animals*; Boas, *Happy Beast*; Hastings, *Man and Beast*; Rosenfield, *From Beast-Machine to Man-Machine*.

viewed and discussed animals was in many ways similar to the way they are perceived today. In this as in so many other respects, the long eighteenth century proved to be the transition to modernity. A major topic of interest among scholars of this topic has been the difference between early modern theriophily ("love of animals"), most notably as exemplified by Michel de Montaigne, and the more stringent Cartesian "beast-machine" theory of animal automatism. The latter has continued for many years to evoke controversy as to the exact level of commitment which Descartes himself had to denying animals sentience.[3] A large number of studies have been written about these topics, and therefore in what follows they will not be discussed in detail, beyond certain necessary references. It should be noted at the outset, however, that despite the seeming disparity between the theriophilic and the Cartesian outlooks, both positions accepted the basic assumption of human superiority to animals, whether in degree or in kind. The clash between animal advocates and their rivals, which has become so conspicuous in our contemporary culture, was much less evident before the nineteenth century. Nevertheless, in the following chapters it should be kept in mind that the varying positions of Montaigne and Descartes were familiar to most Enlightenment intellectuals.

That being said, in the eighteenth century itself these earlier considerations of animals became gradually irrelevant. As we will see in the following chapters, the traditional interest centering on the sensory and mental differences between human beings and animals, which was shared by both the theriophiles and the Cartesians and indeed almost anyone in the early modern era interested in animals, still persisted even in the late eighteenth century. Yet toward the end of the century new modes of discussion of animals, mainly historiographical and economic, gradually displaced this traditional discourse. Descartes's view of animals, which had been so famous, and often notorious, in the seventeenth century, was in fact practically discredited by the turn of the eighteenth century. At the same time the

3 For Montaigne, see Fudge, "Two Ethics"; Panichi, "Montaigne and Animal Ethics"; Boas, *Happy Beast*, passim; Melehy, "Silencing the Animals"; and Shannon, *Accommodated Animal*, 11-17, 183-97 (the last two items comparing him with Descartes). The debate on Descartes began in the nineteenth century and continues today. See e.g. Huxley, "On the Hypothesis that Animals are Automata"; Balz, "Cartesian Doctrine"; Spink, *French Free-Thought*, 226-37; Shugg, "Cartesian Beast-Machine"; Cottingham, "Brute to the Brutes?"; Harrison, "Descartes on Animals"; Steiner, *Anthropocentrism and Its Discontents*, 132-52; Newman, "Unmasking Descartes's Case"; Radner and Radner, *Animal Consciousness*; Senior, "Souls of Men and Beasts"; Rosenfield, *From Beast-Machine to Man-Machine*, passim; Fudge, *Brutal Reasoning*, passim; Thomas, "Does Descartes Deny Consciousness to Animals?"; Friedland, "Friends for Dinner" (also on Montaigne); Strickland, "God's Creatures?"; Miller, "Descartes on Animals Revisited."

theriophilic tradition, which due to its roots in the thought of Plutarch and other classical philosophers, had been quite popular in the Renaissance, was gradually replaced by more modern versions of ethical arguments in favor of sensitivity to animal sentience.[4] In what follows, therefore, the use of the term "theriophily" will often be somewhat anachronistic. It should also be remembered that it was not in common use in the early modern era. It will therefore be used here loosely to designate a wide array of philosophical claims which to varying degrees emphasized animal sentience and usually some level, even if minimal, of consequent ethical obligation to animals. Eighteenth-century theriophiles were no longer interested in the classical exemplars of animal sagacity which had interested Montaigne and other early modern primitivists as part of their critique of human pride. Yet the philosophical difficulty of outlining a theory regarding the differences between human and animal characteristics, and the relevance this had for the ethical treatment of animals, if anything, became of even more wide-ranging interest. It will become apparent that a large majority of eighteenth-century intellectuals shared some level of theriophilic views and that this more often than not did not entail any kind of belief in animal rights in the modern sense. Put briefly, demonstrating some, even token, sensitivity to animal sentience became in the eighteenth century part of a civilized façade, almost a requirement of politesse and respectability. For someone to espouse a Cartesian view of animals, or total lack of sensitivity to their suffering, would have seemed not only philosophically unsound but also practically uncivilized. Yet this did not by any means entail a broadly-shared commitment to extolling animal mental capabilities or to improving their treatment in any significant way. The attempt by various historians to claim the eighteenth century as the historical moment of the roots of the modern conception of animal rights therefore seems to rest on shaky ground. At most, it refers to a phenomenon which was peripheral to the mainstream culture of the period.

One related eighteenth-century development that has often been discussed by scholars is the seeming rise in romantic sensitivity to the aesthetic qualities of nature, which has been linked to the view that the roots of the modern humanitarian consideration of animals originated in the Enlightenment. However, modern scholars ranging from Norbert Elias to Keith Thomas have noted, in different ways, how this emotional and aesthetic sensitivity to nature was the product of urban élites inhabiting

4 For the classical roots of the theriophilic tradition, see Gill, "Theriophily in Antiquity"; and Lovejoy and Boas, *Primitivism and Related Ideas*, 389-420 and passim.

increasingly industrialized cities and hence being ever more removed from direct contact with nature.[5] It was precisely this increasing remove from nature and animals which enabled the emergence of modern environmentalism in general and animal-rights advocacy in particular. This was a type of luxury which less industrialized societies could not afford, a point which remains increasingly evident on a global scale today. Several scholars have noted, in different ways, how this dialectic developed specifically in relation to animals.[6] Of these, the most controversial has been Donna Landry, who has used this point as a justification for hunting in the traditional English countryside fashion.[7] It should, in any case, be kept in mind that the late eighteenth-century changes in conceptions of animals which will be charted below were intimately related to this transformation of European society's relationship with the natural environment.

All this begs the question: if the Enlightenment did not after all contribute in any straightforward way to a clear rise of an unambiguous humanitarian concern for animals (which raises a whole host of philosophical questions, well beyond the confines of the present discussion, regarding what exactly such a humanitarian view of animals might be), and if it did not contribute anything truly novel regarding the traditional debate about animal mental characteristics, what, if anything, was its innovative contribution to the rise of the modern view of animals? The argument developed in the following pages is that there was such a contribution and that it was constituted first by an increasingly historiographical consideration of animals, and the place their utilization played in human cultural progress, and second by an economic consideration which took this historical analytical view and transposed it into the prescriptive realm of nascent modern political economy. This type of detailed historiographical and economic discussion

5 Elias, *Civilizing Process*, 496-7; Thomas, *Man and the Natural World*, 181-91, 300-3; Charlton, *New Images of the Natural in France*, 30-4, 199-220; Porter, "Urban and the Rustic in Enlightenment London"; and Harman, *Culture of Nature in Britain*, 5-6, 344. For the American scene, see Marx, *Machine in the Garden*; and Nash, *Wilderness and the American Mind*. For an interesting application of Elias's ideas to the study of early modern attitudes toward animals, see Sahlins, "Royal Menageries of Louis XIV."

6 Berger, "Why Look at Animals?"; Ritvo, *Animal Estate*, 1-6 and passim; Raber, "From Sheep to Meat." Also of interest are Bradie, "Moral Status of Animals"; Harwood, *Love for Animals*, 64, 74-5, 126 and passim; Hastings, *Man and Beast*, 16, 279-82; Kerestman, "Breaking the Shackles of the Great Chain of Being"; Maehle, "Cruelty and Kindness to the 'Brute Creation'"; and in particular Thomas, *Man and the Natural World*, 181-3, 300-3.

7 Landry, *Invention of the Countryside*. Landry's otherwise sophisticated argument does not sufficiently tackle the question of the inherent moral problem concerned with the enjoyment hunters derive from killing animals.

of animals was barely noticeable before the eighteenth century, yet it subsequently became a mainstay of the modern view of them and their relations with human culture.

The following chapters will outline this changing view of animals throughout the long eighteenth century. The chapters in the first part will concentrate on the seventeenth-century practice of animal experimentation, or vivisection, to use the modern term. This will serve as a prelude to the connection between philosophical and other types of consideration of animals, and to subsequent developments in eighteenth-century discussions of them. The first chapter will raise some important connections between early modern experimentation on animals and the general contemporaneous ethical debate about them. The second chapter will take a closer look at one specific example of this connection, that of the famous Dutch scientist Christiaan Huygens. These chapters will also set the stage for the pan-European nature of the developments charted here, though with particular attention devoted to several countries, not least Holland. The chapters in the subsequent second part will discuss various eighteenth-century examples of literary and philosophical discussions of animals and will convey the broad array of Enlightenment theriophilic positions, many of which were very moderate in their conceptual and ethical implications. Toward the end of this part we will begin to see the transition to a more modern historiographical discussion of animals. The third chapter examines one of the most emphatic, yet least-discussed, examples of theriophilic philosophy in early modern literary culture, the popular epistolary work known as the *Turkish Spy*. This is followed in the fourth chapter by a discussion of perhaps the most famous case of Enlightenment theriophily, that of Jean-Jacques Rousseau; but this discussion will also emphasize the clear limits to Rousseau's theriophily, which are usually not sufficiently recognized by scholars. The next chapter again returns to an almost forgotten case of Enlightenment theriophily, that of the Scotsman William Smellie, visiting the connection between consideration of animals and the wider issue, and limits, of eighteenth-century anthropocentrism and its critique. This is also the first chapter which emphasizes the important role which the Scottish Enlightenment played in developing novel considerations of animals, again demonstrating the pan-European nature of Enlightenment attitudes toward animals. Chapter 6 continues the investigation of Scottish Enlightenment views of animals, giving a general overview of this topic while taking a particular look at another nearly forgotten figure, the physician John Gregory, a much less innovative thinker when it came to animals compared to Smellie but, in consequence, one who was probably more representative of common contemporaneous viewpoints. Chapter 7 then begins the shift, crucial to the argument of this book and mirroring the shift which occurred in Enlightenment thought itself, from more traditional discussions of animal mental characteristics vis-à-vis human beings, and the possible ethical ramifications of these characteristics, to the novel historiographical and economic view of animals. In this chapter the emphasis will be more on the ethical debate about animals as this appeared in the context of Enlightenment natural philosophy, which was to influence the view of nature and animals in Enlightenment historiography. This sets the stage for the final chapter of this part, which will consider in detail the rise of the Enlightenment's historiographical consideration of animals in the work of historians (broadly defined) such as Johann Gottfried Herder and the Abbé Raynal, and (more strictly defined), William Robertson, and most significantly, Edward Gibbon. This chapter will set the stage for the chapters of the third and final part, which will describe the conclusion of this changing conception of animals as it was transposed from the historiographical to the economic field of inquiry. It will do so, somewhat surprisingly at first, by looking at the depiction of animals not just in the obvious realm of political-economic discourse but also in art, specifically in seventeenth-century Dutch painting, thus re-emphasizing the significance of the Dutch contribution to changing conceptions of animals during the long eighteenth century. Joining a discussion of art and economic thought might seem methodologically unconventional, but the logic for doing so will be explained, and it will serve to emphasize how the earlier popular mode of discussing animals, most notably in philosophical discussions with varying levels of theriophilic overtones, was no longer at the forefront of intellectual innovation in the late Enlightenment or, indeed, perhaps not even a century earlier. Chapter 9 discusses the popular early modern genre of paintings of dead animals, and how these relate to contemporaneous philosophical debates about animals. Chapter 10 then moves to a detailed consideration of the most innovative economic discussion of animals in the Enlightenment, not surprisingly that developed by the Scottish philosopher and father of modern economic discourse, Adam Smith. The final chapter then joins all these threads – discussing the emergence of the modern economic view of animals both in art, again specifically Dutch seventeenth-century painting, and in economic literature itself. Some examples of how this new economic consideration of animals was perpetuated in nineteenth- and twentieth-century economic literature will then be surveyed to demonstrate how, by the late eighteenth century, the stage was set for the modern type of economic perception of animals and their place in human civilization, an outlook which is still with us today.

The following chapters do not claim to present a conclusive and comprehensive discussion of eighteenth-century attitudes toward animals. They offer an interpretation of this topic, and they do so while discussing sources and materials most of which have not been discussed in this way before. In addition to philosophical, scientific, and literary materials, which often figure in intellectual history, significant attention will be given to artistic, historiographical, and economic sources, the latter two in particular heretofore having been conspicuously absent from studies of the history of attitudes toward animals. Also, as already mentioned, in addition to English, French, and other sources, particular attention will be devoted to seventeenth-century Holland and to eighteenth-century Scotland. Anyone even remotely familiar with the history of the Enlightenment, its sources, development, and influence, knows that the significance of these two centers of historical cultural innovation requires no introduction. The result of these methodological, thematic, and interpretative points will, hopefully, shed new light on a topic which in recent years has gained increasing scholarly attention.

This leads us to one final point which should be kept in mind throughout the following pages. The study of the history of attitudes toward animals is currently in a transitional state – on the one hand it is no longer a relatively new field of inquiry as it was thirty years ago; but on the other, it is not yet an established field with its own core of methodological and interpretative assumptions shared by most scholars. Indeed, there is not yet even any major point of contention over which scholars are engaged in heated debate (a sure sign that a field of inquiry has become truly significant and of wide interest). A further sign of this unhelpful esotericism is the fact that no major connections have been elaborated between the issues which historians of animals study, and the wide-ranging debate on the nature of the Enlightenment which has been ongoing in the past generation, and is still heatedly contested.[8] As an opening salvo I want to offer here a few general thoughts.

About thirty years ago historians began contesting the traditional view, represented by the impressive overviews of scholars such as Ernst Cassirer and Peter Gay, of the Enlightenment as a unified philosophical

8 For a few good surveys and commentaries, see Robertson, *Case for the Enlightenment*, 1-51; Oz-Salzberger, "New Approaches"; McMahon, "What are Enlightenments?";and O'Brien, "Return of the Enlightenment." Also see Wolloch, "Natural Disasters."

outlook dominating the eighteenth century and the transition from the early modern to the modern world.[9] Increasingly, the emphasis has been on pluralizing the Enlightenment, mainly along national lines but also while emphasizing different intellectual, social, and religious contexts.[10] J. G. A. Pocock has been particularly influential in pluralizing the Enlightenment, claiming that "There is no single or unifiable phenomenon describable as 'the Enlightenment,' but it is the definite article rather than the noun which is to be avoided." Pocock regards reference to *the* Enlightenment as an unavoidable but regrettable reification, and using qualifying adjectives is a reminder "that the keyword 'Enlightenment' is ours to use and should not master us."[11] This tendency, to various extents and in a myriad of ways, to contextualize and pluralize the Enlightenment, has in general predominated eighteenth-century studies for the past generation, not least because it validates the most esoteric of studies. It has added many new perspectives to our understanding of the long eighteenth century, yet it has tended to unjustifiably fragmentize the idea of the Enlightenment as the single most important intellectual, ideological, and ultimately social and political, force behind the great innovations which western civilization underwent during this transformational era.

Some scholars, such as John Robertson, while taking into consideration the insights offered by the pluralizing view of the Enlightenment, have nonetheless insisted on an updated interpretation of the Enlightenment as a more or less unified phenomenon.[12] The most prominent of these has been Jonathan Israel, in several voluminous, influential, and, in some scholars' views, controversial, books.[13] While Israel does not view the Enlightenment as strictly a unified movement, he does regard it as a pan-European one spanning the long eighteenth century. The differences within the Enlightenment itself (discounting the Counter Enlightenment, which was obviously opposed to it) were not along national or other contextual lines but, rather, according to the level of commitment to Enlightenment ideals, and most significantly to the idea of democracy. In accordance with this, Israel differentiates

9 Gay, *Enlightenment*; Cassirer, *Philosophy of the Enlightenment*.

10 A particularly influential volume in this respect has been Porter and Teich, eds., *Enlightenment in National Context*.

11 Pocock, "Historiography and Enlightenment," 83-4, 91, 93-5. Also see Pocock, *Religion: The First Triumph*, 215-19, 313-14.

12 Robertson, *Case for the Enlightenment*. For a different approach, reaffirming France as the center of the Enlightenment, see Edelstein, *Enlightenment: A Genealogy*.

13 Israel, *Radical Enlightenment*; Israel, *Enlightenment Contested*; Israel, *Revolution of the Mind*; Israel, *Democratic Enlightenment*; Israel, *Revolutionary Ideas*.

between the Moderate and the Radical Enlightenments. It is the latter which he sees as evincing the true Enlightenment, as it was manifested in the initial stages of the French Revolution before it was corrupted by Robespierre and his ilk, whom Israel regards as enemies of the Enlightenment. Furthermore, Israel sees the roots of the Radical Enlightenment in the Spinozism of seventeenth-century Holland. It is beyond the scope of our discussion to go in detail into the many criticisms which have been leveled at Israel's interpretation, particularly regarding the Spinozistic roots of the Enlightenment, but also against his generalizing claims about its nature and its role in the revolutions of the late eighteenth century. Some of these criticisms, particularly on specific points, no doubt have merit. Yet such a wide-ranging overview of the intellectual history of the Enlightenment has been offered by no one else, not even by Peter Gay, and it seems petty to look for specific faults and imprecisions, which are bound to occur in such a wide-ranging survey. If nothing else, Israel's work has made it once more legitimate to seriously consider the Enlightenment in a general way, and this is a welcome correction to many years of studies which have become esoteric even among scholars of the eighteenth century. For my own part, if there is one point on which I disagree with Israel, it is that I view the differences between the Moderate and Radical Enlightenments as less emphatic than he does, and therefore the Enlightenment as even more unified.[14]

How does the history of attitudes toward animals figure into this whole debate? No doubt this seems precisely one of those fields which are esoteric, or at least highly specialized, to an extent which makes it the province of specialists, not of those interested in the Enlightenment in general. This, no doubt, is to the detriment of scholars from both ends of this spectrum. In any event, the initial tendency would be to view the rise of a seemingly greater, more modern, moral sensitivity to animals as part of the generally democratizing current of the Radical Enlightenment. This, however, would be a mistake. A radical figure like the Scotsman John Oswald could indeed couple an extreme advocacy of animal rights with a commitment to political revolutionary principles, dying in battle in the revolutionary army in France.[15] Yet Oswald was a rare anomaly. As will become clear throughout the following pages, recognition of animal sentience, and of at least some level of ethical consideration of animals, became a commonplace in the eighteenth century. This would suggest that the Moderate Enlightenment, rather than the Radical Enlightenment, was perhaps the main vehicle for

14 Wolloch, "Natural Disasters."
15 On Oswald see chapter 6 below; and Wolloch, *Subjugated Animals*, 62-3, 126.

effective long-term changes and improvements in the treatment of animals. Furthermore, as already noted, this seemingly new sensitivity to animals was itself a manifestation of a growing distance from the natural world and thus, ironically, of a new type of human control of this world.

Spinoza himself, as I have argued elsewhere, was in fact surprisingly insensitive to the need for an ethical consideration of animals.[16] Therefore, if one were to insist on the Spinozistic origins of the Radical Enlightenment and on the rise of an ethical consideration of animals as part of the Radical Enlightenment, it would be necessary to differentiate between Spinoza himself and subsequent Spinozistic philosophy (in itself a perfectly possible differentiation) and to document the connection between the Radical Enlightenment and discussions of animals. The latter is a much more difficult task, though perhaps not an impossible one. From a different perspective, the following pages will emphasize that many innovations in the consideration of animals during the long eighteenth century were developed initially, if not always in elaborate form, in seventeenth-century Holland, not least in the art of the Dutch Golden Age. This would tend to corroborate the place of Holland in the emergence of Enlightenment culture as it relates to animals, if not specifically emphasizing Spinoza. More broadly, the significant attention which developments in seventeenth-century Holland and eighteenth-century Scotland receive in the following pages underscores the unified view of the Enlightenment as both a pan-European phenomenon, and one which straddled both ends of the long eighteenth century.

Again, how this interacts with any clear view of progress in the human treatment of animals is a more complicated issue. Our unhistorical tendency from a presentist perspective is to view improvements in the treatment of animals in the nineteenth century, for example in the rise of various societies for the prevention of cruelty to animals, as a clear sign of such progress, and the natural assumption is to locate their ideological origins in earlier Enlightenment thought. Yet, even given such progress, it was intertwined with increasing industrial utilization of animals, also the product of Enlightenment economic and scientific progress. The latter point might initially offer the possibility of a critical view of the Enlightenment and its seemingly insensitive domination of nature and animals. Over the years many philosophers, and not a few historians, have succumbed to this unhistorical and erroneous view of the Enlightenment, whether specifically regarding its attitude toward nature or, from an even more general perspective, severely critiquing the so-called "dialectic of the

16 Ibid., 39-44.

Enlightenment." However, if we limit ourselves to animals for a moment, such a view is even more unsubstantiated than the initial urge to connect the proclaimed improvement in the treatment of animals to eighteenth-century developments and the Radical Enlightenment. The Enlightenment was the source of most of the social and political progress we have made in the last three centuries, yet the story of its influence has not been a linear or simple one, and this is particularly true regarding the development of the treatment of animals.

Interestingly, therefore, a consideration of Enlightenment views of animals can contribute to our understanding of the Enlightenment in general, and of its considerations of more oft-studied human topics in the cultural, social, economic, and even political realms. The immediate tendency of trying to consider eighteenth-century attitudes toward animals would be to begin with the more well-known historical studies and debates of Enlightenment approaches to human issues in all these fields and to extrapolate from these to animal issues. However, this would tend to create problems of expectation – for example of "animal developments" comparable to the rise of new approaches to politics, whether these refer to the Radical Enlightenment's espousal of early ideas regarding universal suffrage (from which the Moderate Enlightenment by and large shrank) or to the actual revolutions of the late eighteenth century. How, if at all, such comparisons can be made based on serious historical evidence is no easy challenge. But to go even further, we might ask whether the opposite inference might also be possible – in other words, to begin with the study of animals and try to extrapolate from this to a novel perspective on traditional historiographical debates about human topics in the eighteenth century. The complicated relationship between the growing ethical sensitivity to animals and the growing distance of urban culture from nature, which, again, was relevant to nature in general, might, for example, pose interesting questions and qualifications regarding any simple tale of progress regarding human, or animal, rights. The fact is that the "Whiggish" view of progress, which in my opinion remains generally valid, can only gain in depth and sophistication by confronting such issues and complications. At this early stage of my own thinking about this problem, the only solution I can see on the horizon is to somehow view the development of attitudes toward animals in the long eighteenth century as predominantly part of the Moderate Enlightenment. On the one hand this would explain the moderate rather than radical limits to the modern sensitivity to animals which developed during this period, and on the other hand it will at the same time explain how this sensitivity, qua its limited nature, was able to gain a lasting foothold in modern culture. Parallels with the influence of the Moderate Enlightenment on improvements in human social and political rights would suggest a similar historical trajectory – one of ineluctable, but far from smooth, progress which, contrary to the initial expectations of Moderate Enlightenment intellectuals, eventually led to more radical outcomes than they intended; outcomes, in fact, which were in large measure the consequences of their own ideas. As for scholarly substantiation for such a wide-ranging claim, all I can offer is the evidence presented in the following pages. While I have not made a point of explicitly noting the connection between this broad interpretation and the specific topics discussed in these pages, reading them with this in mind should offer clear substantiation for this interpretation.

All this, of course, is only by way of a general overview of a scholarly and intellectual terrain of which much remains to be studied. Yet if historians of attitudes toward animals wish to break out of the confines of esotericism, particularly vis-à-vis general Enlightenment studies, they have no choice but to begin to confront such questions.

Part I

Animal Experimentation

1 Animal Experimentation and Ethics in the Early Modern Era

The use of animals in scientific experimentation, or, to use the modern term, vivisection, is far from a modern phenomenon.[1] With the rise of modern empirical experimental methodology in the early modern era, it became increasingly common and by the seventeenth century was consistently practiced by anatomists with varying medical, anatomical, and zoological interests. Anita Guerrini's recent detailed history of animal experimentation in seventeenth-century France mirrors a scientific praxis common, to various degrees, in several European countries at the time.[2] In what follows, however, the scientific aspects of animal experimentation will not interest us in themselves, and we will concentrate only on the ethical dilemmas which this posed for at least some early modern scientists. At a time when such experimentation was done without the use of effective anesthetics, the suffering caused to animals was considerable and, for those conducting the experiments, quite conspicuous. The fact that some of the scientists struggled with the moral complications this entailed offers a good vantage-point from which to begin discussing the consideration of animals in pre-Enlightenment Europe.

Animal experimentation, which had become relatively uncommon in the Middle Ages compared with the experiments conducted in antiquity, was conducted in the Renaissance on an increasingly wide scale. The Catholic Church treated the matter in a manner similar to stoic philosophy, supporting the claim that animals, due to their difference from human beings, were devoid of the right to protection from being experimented upon.[3] One of the most prominent Renaissance scientists who performed vivisections in sixteenth-century Italy was Andreas Vesalius. He was one of the first early modern scientists who recognized the suffering that vivisection caused to animals, yet he accorded it no particular significance, writing that there was very little to learn from vivisection of the brain, since "whether we like it or not, but merely out of consideration for our native theologians, we must

1 For a general overview, see Guerrini, *Experimenting with Humans and Animals*; and also Rupke, ed., *Vivisection in Historical Perspective*. Also see Guerrini and Meli, eds., "Issue on Pre-Modern Animal Experimentation."

2 Guerrini, *Courtiers' Anatomists*.

3 Maehle and Tröhler, "Animal Experimentation," 16-18.

deprive brute creatures of reason and thought, although their structure is the same as that of man." And he continued to note that had he not been at personal risk from religious authorities, he indeed would have performed such experiments on animal brains.[4] This tension between the similarities and dissimilarities of animal versus human physiologies has remained a consistent complication, both scientific and ethical, in the history of animal experimentation. Without a minimal similarity, the comparison of animals with humans seems devoid of scientific significance, particularly from the human-medical perspective, yet the more this similarity becomes evident, the more animal experimentation seems unethical. Vesalius's case demonstrates how this similarity appeared sacreligious, from the traditional religious viewpoint, because it threatened human cosmological superiority. Like early modern empirical science in general, animal experimentation, even if unintentionally, was to undermine this traditional cosmological outlook.

Vesalius's student and heir as the chair of anatomy in Padua, Realdo Colombo, reported that high-ranking clergy enjoyed his public experiments, especially one in which he removed the fetus from the belly of a pregnant dog. As he hurt the fetus, the dog began barking, but the moment he put it close to her face she licked it, oblivious to her own pain, going into a rage if something else was put in front of her. The clergy regarded this as an example of the Virgin's love for her son Jesus, while Colombo, more cognizant of the animal's suffering, nonetheless justified it as the price for knowledge. The "poor dog" was for him "rather the happy dog," since she proved the means for acquiring knowledge.[5]

Vesalius and Colombo were among the animal experimenters who established a tradition which was to become even more widespread in seventeenth-century Italy. In 1660, for example, the Tuscan physician Francesco Redi performed experiments in which he let such animals as cockerels, tortoises, foxes, and rams be bitten to death by poisonous snakes, and he compared their death, while performing autopsies, with the state of the corpses of animals put to death in other ways, for example a strangled cockerel, or a ram drowned by holding its head under water.[6] These experiments emphasize the caution of equating animal experimentation with the modern term "vivisection," since the latter refers specifically to live

4 Vesalius, "Last Chapter," 10.

5 Maehle and Tröhler, "Animal Experimentation," 17-19. For more on Renaissance vivisection, see French, *Dissection and Vivisection*.

6 Redi, *Francesco Redi on Vipers*, xv and passim.

dissections, whereas animal experimentation in general was and is much more varied. Straightforward vivisections, however, were quite common. Giuseppe Zambeccari, who ended his life as a professor of anatomy in Pisa, worked as an apprentice to Redi in Florence in the late seventeenth century, doing most of his experimental work under the latter's influence. Zambeccari's vivisections included such things as splenectomies and unilateral nephrectomies in dogs.[7] In a description of experiments conducted by himself and others, he referred to a dog which was subjected to experimentation on a number of occasions, noting that "we began to compare the dog facetiously to the famous and brave Marshal of Rantzau, who through many wounds received in various battles had gloriously lost almost half his limbs."[8] The Danish general Josias von Rantzau, who became Marshal of France, fought mainly in the Thirty Years' War, and during his career was wounded more than sixty times, losing an eye, an ear, a leg, and an arm, eventually dying in a French prison after being arrested at the suggestion of Cardinal Mazarin. The comparison of him with the vivisected dog was apt, although the Marshal of Rantzau willingly accepted his dangerous role. In any case, Zambeccari's approach, referring to vivisected animals as martyrs of science, was one justification developed by scientists aware of the suffering involved in their experiments.

Seventeenth-century animal experimentation was not confined to Italy, and neither did it go unchallenged in scientific circles. An important example was the conservative French anatomist Jean Riolan II, who like William Harvey, whose theory of circulation he criticized, was one of the leading figures in seventeenth-century anatomical practice.[9] Riolan emphasized the anatomical differences between humans and animals. An animal dying under experimentation, which involved unnatural conditions, exhibited reactions possibly very different from those in a healthy human being. Nevertheless, his attitude toward animal experimentation was ambivalent. At least in his early years he engaged in vivisections, despite regarding their medical relevance as limited. Yet he was convinced that significant questions regarding life should not be decided by viewing tortured and dying animals, and he regarded vivisectors in general as butchers rather than physicians. If they became used to animal experimentation, he feared, they would not hesitate to open morbid human patients to satisfy their curiosity.[10]

7 Jarcho, "Giuseppe Zambeccari."
8 Zambeccari, "Experiments," 311-25, the quotation at 319.
9 On Riolan see Guerrini, *Courtiers' Anatomists*, 25-38, 70-2, 64-5, 77-8; and Mani, "Jean Riolan."
10 Mani, "Jean Riolan," 127-8, 139-41; Maehle and Tröhler, "Animal Experimentation," 21-2.

William Harvey, in contrast, was opposed to Riolan not only on the issue of circulation but also regarding his own more straightforward embrace of the animal experimentation praxis. Harvey's experiments included observations of the movements of the heart in different animals ranging from frogs, snails, and fish to dogs and pigs. He criticized those who criticized him for performing animal experiments, and he claimed that scientists relying only on observation of dead human bodies were in fact attempting to construct a universal syllogism based on a universal proposition, something which he considered erroneous.[11] The conceptual challenge which the similarity between humans and animals posed for animal experimentation was particularly evident in early experiments of blood transfusion, which Harvey's discovery made possible. It is no accident that these were conducted mainly on mentally unbalanced people.[12] The German Georg van Wahrendorff pioneered these experiments in 1642 by injecting wine into hounds, and some years later Sir Christopher Wren injected wine and beer into the veins of a consequently inebriated dog. The English Royal Society began transfusion experiments between dogs in 1665, and Richard Lower was particularly active in this field, among other experiments transfusing the blood of a sheep to a man whose character was to be mitigated since his brain was considered "a little too warm." The man survived the operation. Transfusions were also performed in France and Holland, but when one of the patients died after being transfused with the blood of a sheep in an experiment conducted by Jean-Baptiste Denis, such experiments were almost completely discontinued for a century and a half.[13] Nevertheless, transfusion accentuated both the similarities and the dissimilarities between humans and animals, which animal experimentation made ineluctable. Animal experimentation was therefore a cosmological challenge – its scientific viability relied on physiological proximity between humans and animals, while this very proximity challenged human singularity. It is no surprise that early modern philosophers were so interested in providing a mental differentiation between humans and animals. Yet this type of reasoning was to become increasingly secular and scientific and, as we will see, was to be replaced by the end of the eighteenth century with other types of argumentation for human dominance, which relied more on historiographical and economic thinking, and took both animal sentience and human superiority for granted.

11 Harvey, *Circulation of the Blood*, 23, 26, 44, 57, 141-2.

12 Guerrini, "Ethics of Animal Experimentation," 403-4.

13 Sahlins, "Beast Within"; Maluf, "History of Blood Transfusion," 60-8, 103-4; Shugg, "Humanitarian Attitudes," 234.

In any event, transfusion was a relatively mild form of experiment in terms of the suffering caused to animals and often was more dangerous for the human recipients of the procedure. Other types of experiments posed greater moral dilemmas regarding this suffering. In England the prominent scientists Robert Hooke and Robert Boyle were particularly active in these types of experiments, such as fatal experiments with animals placed in air pumps, also known as "pneumatick engines."[14] After witnessing an experiment of Hooke's in which a dog's thorax was opened and the animal kept alive by bellows blowing air into its lungs, the diarist John Evelyn wrote that it "was an experiment of more cruelty than pleased me." Hooke himself was well-aware of the pain this type of experiment caused and was sorry there was no opiate which could prevent it.[15] These types of moral reservations, it should be noted, however, were usually not enough to deter scientists from conducting experiments. The emphasis on the word "more" in Evelyn's statement is relevant here. He himself witnessed other types of experiments and perhaps even participated in some of them, including a blood transfusion from a sheep to a dog, as well as an air-pump experiment conducted by Boyle.[16] Scholars have disagreed on the nature and degree of the ethical qualms which these English scientists felt regarding the suffering caused by their experiments. Boyle, and even more so Hooke, did not believe in the Cartesian beast-machine theory and therefore clearly recognized this pain, preferring not to repeat cruel experiments on the same animal twice, and Hooke in particular also noted the scientific inaccuracy that this suffering entailed.[17]

The experiments, in any event, continued. During the seventeenth century this did not always mean precisely defined experiments meant to validate specific hypotheses but rather, on many occasions, probing into the animal anatomy, for example by ligating or cutting blood and lymphatic vessels, and noting the results. More and more people became involved in experiments in the early modern era, particularly in the prominent university medical schools of Renaissance Padua and later seventeenth-century Leiden. Students and amateur scientific aspirants performed experiments, at times in private, and the curious public could witness them in public demonstrations. In an age less sensible to both human and animal suffering, there was no public

14 Guerrini, "Ethics of Animal Experimentation," 395-6.

15 Shugg, "Humanitarian Attitudes," 231-2; Evelyn, *Diary*, 221-2.

16 Evelyn, *Diary*, 190-1, 218, 221-2.

17 For different interpretations see MacIntosh, "Animals, Morality and Robert Boyle"; Guerrini, "Ethics of Animal Experimentation," 396-8, 400-2, 406; Shugg, "Humanitarian Attitudes," 227-38; and Thomas, *Man and the Natural World*, 174.

political correctness to censure these public exhibitions of suffering. In Holland, and particularly in Leiden, private experiments were often conducted, while anatomical theatres became the scenes of popular demonstrations of vivisections.[18] Leiden University created comfortable conditions for experiments throughout the seventeenth century. Dogs were the favored subjects of vivisections, and those roaming the streets stood a good chance of falling victim to experiments by medical students.[19] The experiments performed by seventeenth-century Dutch scientists often involved extreme suffering but only rarely elicited moral qualms. The physician and Leiden professor Johannes Walaeus, for example, while still a student had been convinced of Harvey's circulation theory by experiments on dogs conducted by Franciscus Sylvius, and he later conducted similar experiments of his own.[20] Another scientist active in Leiden, Charles Drélincourt the Younger, conducted experiments on large numbers of dogs, in one of them attempting to measure the total blood volume of the animal, apparently by exsanguinating it. The dying animal was subjected to a number of other experiments: for example, the carotid arteries were severed and wax was injected into them.[21]

Attempting to prove Franciscus Sylvius's theory regarding the importance of the pancreas in the digestive process, Regnier de Graaf performed a pancreatic fistula experiment on a dog, repeated many times in Holland and France, in which he opened the trachea of the animal to prevent it from barking. The technique of this experiment was so complicated that it was only repeated in the nineteenth century by Claude Bernard.[22] The technique of silencing animals by severing their trachea was perhaps not uncommon, a way for scientists to quiet the atmosphere of the experimental environment. As part of his work on the female reproductive system, de Graaf performed various experiments, though many of them dissections rather than vivisections, on animals such as rabbits, dogs, and pigs. Similar experiments on the reproductive organs, probably dissections rather than vivisections, were also performed by the famous scientist Anthonie van Leeuwenhoek.[23] Yet another famous scientist, Jan Swammerdam, also experimented on animals, though more in line with de Graaf's stringent procedures. While still a student in Leiden, Swammerdam performed respiratory experiments

18 Rupp, "Matters of Life and Death"; Luyendijk-Elshout, "Introduction," 29-30.

19 Lindeboom, "Dog and Frog."

20 Schouten, "Johannes Walaeus."

21 Lindeboom, "Dog and Frog," 289-90.

22 Ibid., 281, 283.

23 Graaf, "Regnier de Graaf," 129, 134, 165-70; Meyer, "Leeuwenhoek as Experimental Biologist," 119-20.

on dogs. These included observing their respiration by connecting bellows to the trachea and also injecting certain substances to the jugular vein. Robert Hooke, as we have already seen, conducted similar experiments. It seems that at least on certain occasions Swammerdam tied the tracheas to prevent the dogs from making noise. He also pioneered the use of frogs in physiological experiments.[24] When van Leeuwenhoek wrote to Christopher Wren describing some of his experiments on frogs, he described how in one of these experiments he "well-nigh bashed in the frog's head, trying to make it keep quiet," and in another case how a frog, "when it was being cut up, voided its dirt into the dish." Although on at least one occasion he displayed some concern when he returned one of the frogs to its habitat, this was a typical reaction to the plight of the animals, a fortiori when these were less anthropomorphic than dogs.[25]

Swammerdam was also present at the anatomical cabinet of Johannes van Horne, where vivisections formed the cornerstone of the experimental work. Among the scientists he met there was the Danish natural philosopher Nicolas Steno. Steno's mentor was his countryman, the Leiden-trained Thomas Bartholin. Together with his young assistant Michael Leyser, Bartholin experimented on dogs while investigating the lymphatic system. They opened the abdomen of a dog several hours after it was fed. Later they experimented on several other dogs which were first strangled to prevent the infliction of too much suffering. But they then opted to repeat the first procedure on another live dog.[26] This concern for the suffering of the animals, albeit limited, was shared by Steno. At van Horne's anatomical cabinet he and Swammerdam vivisected pregnant dogs in an effort to study the respiration of the embryo. Steno was unhappy with the suffering of the experimented animals, and in a letter to Bartholin he wrote about an experiment he had performed:

> I had to keep that dog alive for three hours, one which had actually had to suffer the whole day under this torture. Unfortunately, one attempt is not sufficient to permit drawing a conclusion with absolute certainty. I will therefore have to do the whole thing over again at the first opportunity, but I must admit that it is not without abhorrence that I torture them with

24 Lindeboom, "Dog and Frog," 284-7; Lindeboom, *Letters of Jan Swammerdam*, 5-6, 9-10. On Swammerdam, see also Cole, *History of Comparative Anatomy*, 270-305.

25 Dobell, *Antony van Leeuwenhoek*, 230-4, quotations at 232, 234. For other experiments on frogs see Meyer, "Leeuwenhoek as Experimental Biologist," 104.

26 Luyendijk-Elshout, "Introduction," 26-7, 34-7.

> such prolonged pain. The Cartesians take great pride in the truth of their philosophical system, but I wish they could convince me as thoroughly as they are themselves convinced of the fact that animals have no souls!![27]

While these moral reservations caused Steno to minimize his vivisection activity, it is unclear whether he discontinued it completely.[28] The Dutch scientist Frederik Ruysch was more decisive and gave up vivisections in disgust, limiting himself to dissection of dead animals.[29] Yet Steno's response, let alone Ruysch's, was rare among seventeenth-century scientists. Toward the end of the century, moral opposition to vivisection became gradually more widespread, yet more in popular circles than in scientific ones. Among the latter, those scientists who were sensitive to animal suffering regarded it, as we have already seen, as a scientific necessity. The debate, however, was often more scientific than ethical, centering on the consequences that the physiological difference between humans and animals had for the validity of the experiments.[30] Experimentation, in any event, continued veritably unabated.[31] In the Dutch case it even migrated across the Atlantic. In Dutch-ruled Brazil the German scientist Georg Markgraf, working for Prince Johann Maurits van Nassau-Siegen while the latter governed the colony, dissected a still-living female three-toed sloth, noting among other things that its heart continued pulsating for half an hour.[32]

Seventeenth-century Europe, and particularly countries where the collective efforts of the Scientific Revolution enhanced anatomical and medical experimentation, such as England, France, Holland, and Italy, was the scene of increasing efforts in the field of animal experimentation. Awareness of the animal suffering this entailed was evidently common, yet in most cases failed to prevent scientists from conducting what were often excruciatingly painful experiments. The Cartesian theory of animal automatism offered one route for conciliating the scientists' consciences. Yet it seems that most of them did not believe in it and were perfectly aware of the suffering their activity caused. Seventeenth-century animal experimentation serves as

27 Quoted in ibid., 35-6, where it is noted that this is an abbreviated quotation. For a similar quotation see also Guerrini, "Ethics of Animal Experimentation," 406.

28 Lindeboom, "Dog and Frog," 64.

29 Lindeboom, *Letters of Jan Swammerdam*, 175; Guerrini, "Ethics of Animal Experimentation," 406.

30 Rosenfield, *From Beast-Machine to Man-Machine*, 88-9; Maehle and Tröhler, "Animal Experimentation," 21-3, 27; Guerrini, "Ethics of Animal Experimentation," 405-7.

31 Guerrini, *Courtiers' Anatomists*, passim.

32 Whitehead, "Georg Markgraf," 440.

an excellent starting point for a discussion of considerations of animals in the long eighteenth century. Few other human activities put in such stark terms the tension between the physiological and mental similarities and dissimilarities between humans and animals, and the ethical complications this entailed. There was also a distinctly modern tone to the scientific discussion and treatment of animals, which was most often devoid of the traditional religious cosmological arguments in favor of human supremacy. Animal bodies were probed because of their strictly physical similarity to human beings, and their feelings and pain were conceptualized in a similarly rationally defined manner. The fact that many of the scientists obviously loved nature as well as animals should also be taken into account. There was a strict difference, however, between loving nature and using it, as will become increasingly clear in the following chapters. A further tension can also be noted between the increasingly rational approach of modern science, and the occasional moral qualms expressed by some scientists, and the parallel development of early Enlightenment political thought, and specifically republican critique of despotic monarchism. The England which saw the experiments noted above was also the England of Hobbes and Locke, and contemporary Holland was also the Holland of Spinoza and, according to Jonathan Israel's interpretation, the most significant originating source of the Radical Enlightenment. Making a direct connection between animal experimentation, with the moral dilemmas it involved, and political discourse is tenuous at best. Yet an indirect, corollary connection has to be noted. The rational and scientific, as well as moral foundations of the Enlightenment came into play in both instances.

2 Christiaan Huygens and Animal Experimentation

Examining in more detail how one prominent seventeenth-century scientist treated animals, both on a philosophical and a practical level, can give a sense of the intricacies of contemporaneous views of animals. The Dutch scientist Christiaan Huygens is known for his invention of the pendulum clock, founding of the wave theory of light, and discovery of the true shape of the rings of Saturn, to name only a few of his prolific activities.[1] Yet he does not figure among the prominent anatomical and medical inquirers of the Scientific Revolution. His consideration of animals, however, was quite intriguing and demonstrates in detail the increasing early modern tension between recognition of animal sentience, on the one hand, and the incentive to utilize animals, and the natural world in general, on the other.

Toward the end of his life, Huygens worked on his last important work, eventually published posthumously in 1698 and usually referred to as the *Cosmotheoros*.[2] In this work he made some short claims on animals which were distinctly opposed to the Cartesian beast-machine theory with its claim for the lack of sentience in animals.[3] According to Huygens, some animals displayed varying levels of rationality. Dogs, apes, beavers, elephants, and some birds and bees, had "somewhat in them of Reason independent on, and prior to all teaching and practice."[4] Therefore human beings were not the only rational animals. "But still no Body can doubt, but

1 On Huygens in general see Bell, *Christian Huygens*, 5-96; Bos, "Christiaan Huygens"; and Bos, "Huygens, Christiaan."

2 The common use of this original title is retained here. The edition used is the first English translation. See Huygens, *Celestial Worlds* [hereafter cited as *CW*]. The work was originally written in French, and various editions and translations were published in the eighteenth century. See Dick, *Plurality of Worlds*, 129-30, 135; Seidengart, "Les théories cosmologiques," 211; Huygens, *Oeuvres Complètes*, 21: 674-5 [hereafter cited as *OCH*].

3 The editors of his complete works have provided a rather cursory explanation of Huygens's views on animals. They explain the differences between his and Descartes's views on animals as a function mainly of their differing religious views and of Huygens's ideas on determinism. See ibid., 21: 662, 667. Yet the attempt to differentiate between Catholic and Protestant views of animals is problematic at the very least. Discussions of the *Cosmotheoros*, in any event, usually disregard the issue of animals. See Bell, *Christian Huygens*, 200-2; Dick, *Plurality of Worlds*, 126-35, 186-7; Knight, "Celestial Worlds Discover'd"; Knight, "Uniformity and Diversity," 67-8; McColley, "Seventeenth-Century Doctrine," 407-9, 416, 423-4; Munitz, "One Universe or Many?" 606-8; Rossi, "Nobility of Man," 146, 155, 158; Seidengart, "Théories cosmologiques."

4 Huygens, *CW*, 56-7.

that the Understanding and Reason of Man is to be prefer'd to theirs [the animals'] as being comprehensive of innumerable things, indued with an infinite memory of what's past, and capable of providing against what's to come."[5] In what related to things such as self-preservation, education, and providing for themselves and their offspring, animals performed most of these things with greater facility than human beings. The latter's sense of virtue, justice, friendship, gratitude, and honesty were meant to put a stop to their own wickedness, or to provide security against mutual assaults and injuries, things in which animals "want no Guide but Nature and Inclination." Therefore, if one compared the many cares, disturbances of mind, restless desires, and dread of death, which resulted from human reason with "that easy, quiet, and harmless Life" which animals enjoyed, then one was "apt to wish a change." Animals, especially birds, enjoyed greater happiness than human beings were capable of, despite all their seeming wisdom.[6]

Human superiority was clear, however, and consisted, according to Huygens, mainly in the human propensity to contemplate the works of God and to study nature, as well as in the improvement of the sciences which fostered knowledge and led to contemplation. The difference between animals and human beings, however, was also mirrored in differences between various humans among themselves, for example between those who looked at the stars and planets with careless supine negligence, compared with those who understood their positions and sizes, or between "such a one as admires perhaps the nimble Activity and strange Motions of some Animals, and one that knows their whole Structure, understands the whole Fabrick and Architecture of their Composition."[7] Huygens was therefore willing to accept points of resemblance between human and animal mental capabilities, without inferring from this any consequent equality. This was a common form of moderate early modern theriophily. It precluded any acceptance of the Cartesian beast-machine theory, on which Huygens was unequivocal, stating that animals

> have as great a gusto of Bodily Pleasures as we, let the new Philosophers say what they will, who would have them go for nothing but Clocks and Engines of Flesh; a thing which Beasts so plainly confute by crying and running away from a Stick, and all other Actions, that I wonder how any one could subscribe to so absurd and cruel an Opinion.[8]

5 Ibid., 57.
6 Ibid., 58-9.
7 Ibid., 60-2.
8 Ibid., 59-60.

The question of Huygens's ambivalent attitude toward Cartesianism is one of the central issues in modern studies of his work. From an almost profound youthful Cartesianism he moved in later years toward a more critical approach.[9] However, acceptance of general Cartesian propositions did not necessarily entail agreement with the beast-machine theory, which was much less popular and was already outmoded by the late seventeenth century. When exactly Huygens formed his opposition to the beast-machine theory, and whether this happened independently or under the influence of others, remains unclear in any event. While still a youth he corresponded with, and was influenced by, the philosopher and scientist Marin Mersenne, who opposed the beast-machine theory, and in Paris he met the elderly philosopher Pierre Gassendi, who also opposed it.[10] Also in Paris he occasionally visited the salon of Madelleine de Scudéry, another opponent of the beast-machine theory, where he may have met proponents of the Gassendist theory of animal soul as a flame-like substance.[11] Another source of influence may have been Leibniz, whom Huygens met for the first time in 1672 in Paris. In a letter from 1684 to his and Spinoza's friend, the German scientist Ehrenfried Walter von Tschirnhaus, Leibniz wrote:

> In Holland they are now disputing, loudly and soundly, whether beasts are machines. People are even amusing themselves by ridiculing the Cartesians for imagining that a dog that is clubbed cries in the same way as a bagpipe which is pressed. As for me, though I grant the Cartesians that all external actions of beasts can be explained mechanically, I nevertheless believe that beasts have some knowledge and that there is something in them, not itself extended, which can be called a soul, or if you prefer, a substantial form.[12]

9 For general discussions of Huygens's Cartesianism, which usually do not discuss the animal issue, see Burch, *Christiaan Huygens*, 144-50, 168-81; Dijksterhuis, *Mechanization of the World Picture*, 457-8; Dugas, "Sur le cartésianisme"; Elzinga, "Christiaan Huygens' Theory"; Elzinga, *Research Program*, 14, 16-18, 27-34, 36-7, 39-40, 80-5; and Westman, "Huygens."

10 Bell, *Christian Huygens*, 25. In a letter from September 1646 to Christiaan's father Constantijn, Mersenne expressed disagreement with the Cartesian assertion that animals were devoid of souls. See Huygens, *OCH*, 1: 21. Also see Gassendi's remarks in his objections to Descartes's *Meditations*, in Descartes, *Philosophical Works*, 2: 144-6 (Descartes's reply to these remarks is in ibid., 211-12); Bell, *Christian Huygens*, 27-8; Boas, *Happy Beast*, 91, 132-5; Rosenfield, *From Beast-Machine to Man-Machine*, 10-11, 13, 25, 110-14, 143, 157, 159, 175-6, 188, 271-2.

11 Ibid., 113, 158-63, 201; Bell, *Christian Huygens*, 95; Boas, *Happy Beast*, 141-2.

12 Leibniz, *Philosophical Papers*, 275-6. For Leibniz's connections with Huygens, see Bell, *Christian Huygens*, 69, 80-1, 88-91, 96, 211. For Leibniz on animals, see Kulstad, "Leibniz"; and Miles, "Leibniz."

This was indicative of the general discrediting of the beast-machine theory by the early stages of the Enlightenment. As for Tschirnhaus, he was Huygens's associate during the 1670s, and one cannot rule out that among other topics they discussed the popular issue of the beast-machine.[13] In any case, Huygens became aware of the animal debate at an early age and at some time formed an anti-Cartesian and moderately theriophilic philosophical position on this issue. His views on animals were connected to his position on the question of the possibility of the existence of extraterrestrial life, a subject which had been discussed since classical times and which was particularly popular during the early modern era, offering in many instances comments which were also relevant to cosmological views of animals.[14] Indeed, Huygens's condemnation of the beast-machine theory was included in the *Cosmotheoros*, one of the most important early modern books on the subject of other populated worlds. The popularity of this subject received additional impetus in the early modern era thanks to the Copernican Revolution, with its demolition of the geocentric image of the universe. Thus the assumption that there might exist other creatures in other worlds which, moreover, might even be as intelligent as human beings, if not more so, gained in popularity. The possibility that such creatures might exist shook the anthropocentric biblical cosmology, and thus in effect there was a correlation between the Copernican Revolution in astronomy and in other fields of natural philosophy.

The anti-anthropocentric aspects of this development were evident in Huygens's *Cosmotheoros*, in which he attacked the religious notion that everything was created for the sake of humanity. In fact, denying the possibility that there might exist other rational creatures on other worlds was an opinion contrary to the "Holy Writ," since there such worlds and creatures were not mentioned and therefore were not denied.[15] It should be emphasized, however, that this type of criticism of human failings, utilizing comparisons to animals, to primitive human societies, or to putative extraterrestrial beings, was ultimately meant not to deny human cosmological supremacy but, rather, to ameliorate human failings, particularly in advanced European societies. This was a common thread running through the European primitivistic and theriophilic traditions. This was no doubt Huygens's position as well.

13 Bell, *Christian Huygens*, 74-5.

14 For a detailed discussion see Wolloch, chapter 3: "Animals and the Early Modern Debate about Extraterrestrial Life," in *Subjugated Animals*, 73-98; and the earlier version of this chapter: Wolloch, "Animals, Extraterrestrial Life and Anthropocentrism."

15 Huygens, *CW*, 6-8.

A belief in extraterrestrial life did not, however, necessarily entail a theriophilic view of animals. The chief evidence for this was provided by Descartes himself, who did not rule out the existence of other worlds and creatures, even though on this, as on the issue of animal sentience, he remained agnostic. His views on possible extraterrestrial life were related mainly to his vortex theory, although he and other Cartesians following him preferred not to emphasize this aspect of Cartesian cosmology, both for theological reasons and because this cosmology was already fraught with problems. Yet Descartes played a key role in undermining the Aristotelian world-view, which together with biblical cosmology had served as the basis for the traditional anthropocentric outlook. This does not mean, however, that Descartes was a primitivist or a theriophile; quite the contrary. His philosophical aim, on this as on other topics, was to provide a new rational argument for religious belief. Therefore only in relation to criticism of limited worldly matters did he evince any notion of anti-anthropocentrism, and he noted that human beings needed also to attend to things beyond the physical world. Therefore his argument for possible extraterrestrial life had no bearing on his views on animals (which were also, one should remember, eventually rather agnostic compared with those of some of his followers) and did not negate the beast-machine theory.[16] The similarity between early modern views regarding extraterrestrial life, on the one hand, and the mental characteristics of animals, on the other, was confined mainly to the potential for anti-anthropocentric argumentation which both offered but which was not necessarily always developed in tandem. Descartes offered one route; Huygens, with his more expansive philosophical view of animals, quite another.

Therefore it is not surprising to find Huygens claiming that reason was a divine element in human beings while at the same time implying that it was impossible that everything in nature was created for their use, even though he approved of this use. He thus combined anti-anthropocentric arguments for a plurality of worlds and rational creatures, together with distinctly anthropocentric views of humanity's utilization of nature, when he wrote:

> That which makes me of this Opinion, that those Worlds are not without such a Creature endued with Reason, is, that otherwise our Earth would have too much the advantage of them, in being the only part of the Universe that could boast of such a Creature so far above, not only Plants and Trees, but all Animals whatsoever: a Creature that has a Divine

16 For Descartes's views on this issue, see particularly his letter to Pierre Chanut from June 6, 1647, in Descartes, *Oeuvres*, 5: 50-8; and Dick, *Plurality of Worlds*, 106-41.

> Somewhat within him, that knows, and understands, and remembers such an innumerable number of things; that deliberates, weighs and judges of the Truth: A Creature upon whose account, and for whose use, whatsoever the Earth brings forth seems to be provided.[17]

That "so noble an Animal," man, converted everything on earth to his own needs. He made houses from trees, stones, and metals, and "the Birds and Fishes he sustains himself with."[18] For Huygens, as for Descartes and most other early modern philosophers, the possibility of the existence of rational extraterrestrial creatures did not revoke humanity's privileged and God-given superiority on earth. Other creatures in far-away planets might be superior to human beings, but the terrestrial animals were definitely not.

Huygens might not have been a particularly interesting or original figure in the history of early modern attitudes toward animals had it not been for the fact that he also conducted experiments on animals, mainly with the air-pump or "pneumatic engine." In comparison with the scientists discussed in the previous chapter, however, Huygens provides an excellent example of how the inclination to conduct cruel experiments on animals could be reconciled with a theriophilic philosophical position. The air-pump had been invented in 1647 by the German scientist Otto von Guericke, who among other experiments observed what happened to animals in a void.[19] Robert Hooke and Robert Boyle, as we have seen, also performed similar experiments. Huygens's main scientific activity was not, of course, in this field. Yet his polymathic interests led him also to conduct similar experiments, mainly during the 1660s and 1670s. He first became interested in the air-pump in 1661 while visiting London. He was interested in Hooke's design for the instrument and Boyle's experiments with it. In the following years he made important contributions both to the design of and to the experimentation with this instrument.[20] He also influenced the use of it by others, for example through his demonstrations before members of the Académie Royale des Sciences in Paris in 1668.[21]

17 Huygens, *CW*, 37-8. See also in general the arguments in *CW*, 7-8, 36-9. Also of interest are the editors' remarks in Huygens, *OCH*, 21: 663-4.

18 Huygens, *CW*, 38.

19 Guericke, *New (So-Called) Magdeburg Experiments*, 143. I have not been able to find expressions of sorrow at animal suffering by von Guericke. But it is interesting to note that he acknowledged the possibility of the existence of extraterrestrial creatures and opposed anthropocentrism.

20 For his activities in this field, but without discussion of his animal experiments, see Stroup, "Christiaan Huygens."

21 Ibid., 139.

What makes Huygens's activity as an experimenter on animals interesting is that, despite his theriophilic views, his descriptions of these experiments were extremely factual and included no substantial reference to the suffering of the animals. Of course the theriophilic views expressed in the *Cosmotheoros* were the product of his last years, long after he conducted his animal experiments. But as we have seen, they might just as well have matured in his mind at any time since his youth. They thus exhibit a cognitive dissonance when superimposed with his notes on the animal experiments. The closest Huygens seems to have come to exhibiting concern for an animal experimented upon was when he related conserving the life of a small bird, which had fainted in the air-pump, by quickly pulling at a part of the machine. Yet even this description was very unemotional and did not seem to exhibit any kind of special concern for the animal's suffering.[22] Huygens's tone in descriptions of such experiments was habitually quite dry and did not disclose the kind of moral qualms that certain other vivisectors expressed. There are several such factual references in his letters and writings. In a letter to his brother Lodewijk from November 30, 1661, he wrote of placing sparrows and mice in an air-pump.[23] It is probably one of these experiments to which he referred in a tract on the air-pump, as he described an experiment made on November 29, 1661, in which a canary was put in an air-pump a few times until it died. The language was unemotional:

> It [the canary] did not yet appear very uncomfortable the first two times, but then it began to pant and after that it leaned its head while blinking its eyes. Then it revived, for an instant was completely alert and flapped its wings; but soon it again became tranquil and fell unconscious, dead. Each time one saw it swell somewhat.[24]

Other references were similarly unemotional, as when he related a few days later, in another letter to Lodewijk, how he put a small bird in an air-pump, which died "just like the one of which Monsieur Boyle related."[25] Boyle's

22 This description is included in a letter by Huygens to the scientist Sir Robert Moray, from January 4, 1662. See Huygens, *OCH*, 4: 8.

23 Ibid., 3: 395.

24 "Il ne parut pas encore très incommodé par les deux premiers coups, mais ensuite il commenca à haleter et après cela il pencha la tête en clignotant des yeux. Puis il se ranima, fut un instant complètement en eveil et battit des ailes; mais bientôt il redevint tranquille et tomba sans connaissance, mort. A chaque coup on le voyait s'enfler quelque peu." Ibid., 17: 312. Translations are mine unless otherwise noted.

25 Letter from December 7, 1661, in ibid., 3: 397.

influence was also evident in Huygens's letter several years later to Henry Oldenburg, secretary of the Royal Society in London, in which he praised Boyle's experiments on respiration and claimed that further experiments should be made to discover the role of air in animal nourishment and how it was transferred to the blood.[26] Other instances further substantiated this point. The experiments made with the air-pump at the Académie Royale des Sciences in 1668 were described in language that makes it unclear whether Huygens himself actually performed them, but it is clear that they were at least conducted under his supervision. The descriptive style is again stark and factual:

> We placed a live mouse in the recipient, which battling strongly the first time, when we pumped the air, appeared very feeble the second time, and the third, remained spread out without movement. We immediately restored the air to it, but it didn't move, and having been pulled out of the recipient, was found dead. Some time afterwards we conducted a dissection of it, and we remarked nothing extraordinary in its body, only the lung seemed a little withered.[27]

A similar experiment was conducted before the Académie several weeks later on a fish, which did not die in the air-pump but was subsequently dissected. Again the description was unemotional.[28] Huygens's assistant Denis Papin published a book on the air-pump which probably described work done by, with, or following Huygens.[29] The descriptions of experiments there were likewise dry and factual.[30] While Huygens performed animal experiments mainly with the air-pump, it is not unlikely that he also performed other

26 Letter from October 31, 1670, in ibid., 7: 44. Huygens's support of biological research was of course not connected only with Boyle's work. In a letter from 1666 to Colbert, probably in connection with the foundation of the Académie Royale des Sciences, he mentioned various important scientific activities, among which he included the study of animal respiration and the dissection of animals. See ibid., 6: 95-6.

27 "On a mis dans le recipient une souris vivante, laquelle s'estant fort debattue la premiere fois, qu'on pompe l'air, parut fort affoiblie la seconde fois, et la troisiesme, demeura étendue et sans mouvement. On luy redonna aussytost de l'air, mais elle ne remua point, et ayant esté tirée hors du Recipient, elle fut trouvée morte. Quelque temps apres on en fit la dissection, et on ne remarqua rien d'extraordinaire dans son corps, si ce n'est que le poumon sembloit estre un peu flestry." Experiment conducted on April 14, 1668. Ibid., 19: 207.

28 Experiment of May 5, 1668. Ibid., 19: 211.

29 Bell, *Christian Huygens*, 215; Stroup, "Christiaan Huygens," 135.

30 The book was published in 1674. The descriptions of the vivisections in it are found in Huygens, *OCH*, 19: 231-3.

types of experiments, or at least that he was not opposed to this practice. In a letter from Paris to Lodewijk in 1664 he related having assisted in a dissection of one of three dogs which had earlier been subjected to a splenectomy. He did not seem to have performed the experiments, but again he displayed no aversion to them.[31]

The seeming inconsistencies within Huygens's attitude toward animals make his views on this subject seem at first sight extraordinary, and therefore historically non-indicative of general developments and trends. But this is not the case. The juxtaposition of seemingly contradictory attitudes toward animals is one of the mainstays of historical attitudes toward them, particularly in the early modern era. The praise of nature and animals, which did not participate in the original sin and were implicated in it only because of human beings, was from biblical times accompanied by the view of humanity as the steward and summit of nature. Human beings were sinful and subjected to constant criticism, but at the same time were also the most perfect terrestrial creatures. Their very capacity for sin, with its obverse capacity for saintliness, differentiated them from the animals. It was this conceptual tension which accounted for much of the intellectual energy diverted throughout the ages to questions regarding humanity's attitude toward nature in general, and toward animals in particular. Animal experimentation before the development of effective anesthetics brought this tension into special focus. Here was a human activity which made the suffering of animals, which most scientists were clearly aware of and did not deny in Cartesian fashion, emphatically tangible. Seen in this light, expressions of concern for animal suffering by scientists could be viewed as signs of embarrassment in the face of this reality. Conversely, lack of such expressions by at least some of them could be viewed as simply a less tortuous acceptance of this tension.

What makes Christiaan Huygens's attitude toward animals interesting in this context is precisely the apparent lack of conflict between his views as a critic of the "absurd and cruel" Cartesian beast-machine theory, on the one hand, and his activity as a scientist seemingly oblivious to animal suffering, on the other. One may of course surmise that he did have qualms regarding the suffering he caused to animals in his experiments, but decisive proof

31 Letter of April 26, 1664, in ibid., 5: 60. In Paris Huygens constantly assisted in dissections of various animals, out of interest in their anatomy. See the editorial comment in the biographical section in ibid., 22: 752. The term "dissection" in seventeenth-century descriptions of experiments may at times have included also procedures done not just on dead animals but on live ones as well, i.e., vivisections. Therefore when Huygens remarked here that he assisted in a dissection it is possible that he was in fact referring to a vivisection, although this cannot be ascertained.

of this is lacking. In any event, this type of approach, as we have already seen, was commonplace among contemporary scientists. Moral scruples had in most cases only a mild influence on their actual experimental activity. It was precisely the matter-of-fact references to animal experimentation depicted by Huygens which exemplified the historically predominant attitude toward the scientific exploitation of animals. Animals may have been praised in laudatory philosophical terms, but when it came to their actual physical utilization, the translation of theoretical arguments into reality was customarily forsaken. Huygens and his colleagues were not oblivious to animal suffering, but they regarded such suffering as worthy of moral consideration only when such consideration did not contest human needs, specifically scientific experimentation. In this respect, even if Huygens would have expressed regret at causing animal suffering, it probably would not have made much difference from the practical point of view of continuing to perform his experiments. His awareness of the feelings of animals expressed in a different context simply accentuated this fact. As we will now see, the willingness to accommodate ethical sensitivity to animals with the more overarching need to utilize them would receive a broader theoretical underpinning in the following century. Animal experimentation continued in the eighteenth century. But in terms of a juxtaposition of these seemingly conflicting outlooks, the Enlightenment was to present a detailed philosophical outlook emphasizing a growing historiographical and economic consideration of animals.

Huygens's activity in Holland and France, two of the most important centers of the early Enlightenment, also reminds us again of the connection between the attitude toward animals in both science and moral discourse and of the broader political concerns of the period. Huygens, despite his criticism of the beast-machine theory, was, like others in the late seventeenth century, deeply indebted to Descartes's rational discourse. Descartes's attempt to provide a new rational argument for religious belief was to have the unintended consequence of ultimately serving the growing Enlightenment critique of this belief, at least in its more traditional forms. The consideration of the possibility of extraterrestrial life also challenged traditional religious cosmology. Even though they often did not intend to do so, seventeenth-century scientists and rationally inclined philosophers were contributing to erect a conceptual outlook which increasingly supported a similar revolutionary approach to politics. Changing views of animals were in this respect at least tangentially related to the broader concerns of the Enlightenment.

Part II

From Philosophy to Historiography in the Enlightenment

3 *The Turkish Spy* and Eighteenth-Century British Theriophily

The early modern debate regarding animal mental characteristics was not confined to philosophers and scientists. Not surprising, it spilled over into fictional literature, which was not only more accessible and widely read but also more inclined to express concern with animal suffering. One example of this literary phenomenon at the turn of the eighteenth century is the subject of the present chapter. *Letters Writ by a Turkish Spy*, usually known simply as *The Turkish Spy*, was a popular work of fiction in the eighteenth century, particularly in England. The first volume, published in French in 1684, was almost certainly written by an Italian, Giovanni P. Marana. As for the authorship of the following seven volumes, published in English between 1691 and 1694, various possibilities have been raised, such as Marana himself, an unknown English author (or authors), or perhaps Protestant exiles in England. The work was written as a collection of letters by Mahmut, a spy for the Turkish Empire, who resided many years in Paris and corresponded with various officials, family, and friends, reporting on the occurrences, fashions, and customs of seventeenth-century Europe, mainly France. Mahmut was of course a fictional protagonist, who voiced observations on late seventeenth-century Europe as if through the eyes of a non-European. Therefore, his was ultimately an occidental point of view. While *The Turkish Spy* was written toward the end of the seventeenth century, its popularity and influence were apparent particularly in the following century. It inaugurated the genre of foreign observer epistolary literature which was to become well known in such works as Montesquieu's *Lettres persanes* and Goldsmiths's *Citizen of the World*. Yet in itself it has received relatively little attention in modern scholarship.[1] In particular, the references to animals in *The Turkish Spy* have received only cursory discussions.[2] *The Turkish Spy*, however, was one of those fictional works which made complicated philosophical ideas accessible to a wide reading public, and therefore the

1 The edition consulted here is the first complete English edition of the work – *Eight Volumes of Letters Writ by a Turkish Spy* [hereafter cited as *TS*]. For general introductions, see the editorial introduction in Weitzman, ed., *Letters*, vii-xix; and Gaudier and Heirwegh, "Jean-Paul Marana."

2 See the short references in Weitzman, "Introduction," passim; Nicklas, "Letters," 360-2; Pope, Arbuthnot, et al., *Memoirs*, 292-3.

theriophilic overtones it contained no doubt both stemmed from and fostered sensitivity to animal sentience.

The many passages on animals in *The Turkish Spy* were written in a consistently theriophilic vein. Mahmut (this "familiar" reference is used here throughout for convenience) continually returned to discussing animals and exhibited an acquaintance with most points of early modern theriophilic argumentation, even if much of this was couched in the form of criticism of European culture, as opposed to what he claimed was a more moral conduct in the Ottoman Muslim world.[3] The issue of animal characteristics received constant attention, and Mahmut claimed that "to understand well the different Natures of Beasts, is a Study fit for Kings."[4] In the style of classical and early modern theriophily, he related many stories of animal sagacity and virtue.[5] He wrote in this way even of insects, for example of the sagacity of bees, whose industry was comparable to human government, or of the spider, which he beheld with pleasure as it built with exquisite artistry its "little silken Palace" with which to catch flies.[6] Animals exhibited many of the qualities, both good and bad, that human beings possessed, even if to a lesser degree. As there were differences between various individual people, or between nations, so did similar differences exist between various animal individuals and species.[7] The comparison with human beings was continually reiterated, for example when Mahmut wrote:

> our [human beings'] Will has little to do in the Conduct of our Lives. We, like all other Creatures, act according to certain Secret Impulses of Nature. The very same Faculty which we call Instinct in the Beasts, is no other than what we term Reason, Wisdom, Knowledge, Discretion and Forecast in our selves. And I think 'tis no Solecism to say, That that was a Prudent Dog, who perceiving his Master making ready a Rope to hang him, slily slipt away, and never came near him more.[8]

3 I am not in a position to verify the nature of early modern Muslim attitudes toward animals. In any event, the comparison of western and pagan views is used in the work principally as a device for criticizing the former.

4 *TS*, vol. 3, book 2, letter 5, 150-2.

5 Ibid., 4, 3, 4: 217-22; 4, 4, 3: 305-10; 4, 4, 7: 328-32.

6 Ibid., 2, 1, 26: 93-6.

7 Ibid., 7, 3, 8: 220-5.

8 Ibid., 5, 4, 1: 277-83, at 279.

He claimed that "All animals seem ... to be endued with a Faculty, which if it may not be called Reason; yet is something Analogous to it, for which we want a proper Name."[9] Mahmut, therefore, did not regard animals as equal to human beings, writing:

> I have been often at a Loss, in Contemplating the Soul of Man. Sometimes it seems no otherwise distinguish'd from the Souls of Brutes, than by being United to a Body of Different Organs; Which causes us to shew more Evident Tokens of Reason than they, in the Faculty of Discourse, and in our Actions. Yet when I consider more attentively the Operations of our Mind and Intellect, I cannot but conclude, there is a vast Distinction between our Souls, and those of the Beasts.[10]

However, if human beings were superior to animals in many things, the latter also had their fields of pre-eminence. "I easily agree, that we far excel the other Living Creatures in all the Operations of our Souls, and Exercises of our Reason: Yet, we have our defects as well as they; and, this I esteem as one of the greatest, to deny them any share in *Reason*, who so far excel us in *Sence* [*sic*]."[11] The claim that animals compensated for their inferiority to human beings in mental abilities by a superiority in sensual capabilities was commonplace in theriophilic literature. Mahmut gave an example.

> I cannot forbear doing this Justice to the Fish of the Sea, as well as to the Animals on Earth, to acknowledge, That either they are indu'd with a Kind of Reason; or, that Faculty which we call so in Men, is no other than Sense. If the Brutes perform many Things without any Deliberation or Counsel, so do most Men: And no Man can demonstrate, That even those Dumb Beings, do not advise and project, before they attempt any Thing of Moment towards their own Preservation, or the Service of others. And if they seem to do many Things rashly, it may be attributed to the Quickness and Vivacity of their Sense, which needs not the Slow and Flegmatick Methods of Human Counsel.[12]

9 Ibid., 2, 1, 26: 93-6, at 94.
10 Ibid., 5, 4, 16: 350-2, at 351.
11 Ibid., 2, 1, 26: 93-6, at 96.
12 Ibid., 4, 3, 19: 282-7, at 284.

Mahmut's writing on animals was clearly influenced by classical philosophy. For example, he wrote of "*Porphyry* the *Philosopher*, who seems to approach nearest to *Reason* of all the *Ancient Sages*."[13] Porphyry, the third-century Neoplatonic philosopher, had written a book on vegetarianism and was familiar to early modern scholars. His arguments for vegetarianism included a belief in metempsychosis, a recognition of the rationality of animals, and the popular claim that the treatment of animals led to similar treatment of human beings.[14] Mahmut, however, was critical enough to note that the claim denying animals reason, which he regarded as erroneous, had also originated in antiquity.[15] His conjectures regarding metempsychosis were particularly influenced by classical philosophy. He was not sure whether transmigration of souls did indeed occur, but he regarded it as a definite possibility, in a clearly implied Pythagorean vein.[16] A further classical influence was evident in the allusion to the theory of the Great Chain of Being, which was particularly popular in the early modern era, with its view of human beings as situated in between the animals and the angels.[17] He was exasperated by his inability to know what happened to the soul after death, and this led him to raise the possibility that the transmigration of souls occurred.[18]

Mahmut consistently compared European treatment of animals unfavorably with non-western cultures. The Muslim, and specifically Turkish, attitude toward animals was continually portrayed approvingly, although the Egyptians, Persians, ancient Greeks, ancient Jews, and even the Druids were also similarly mentioned.[19] Particular attention, alongside the Ottoman world, was given to the benign treatment of animals in India. Mahmut repeatedly returned to this theme, mentioning how in India it

13 Ibid., 5, 4, 9: 315-19, at 319. This remark is not connected to the discussion of animals. But given Mahmut's remarks on vegetarianism, noted below, the importance of the theriophilic aspects of Porphyry's thought seems clearly implied.

14 Porphyry, *Abstinence*. For Porphyry's influence in early modern England, see Thomas, *Man and the Natural World*, 180, 289-97. Also see Steiner, *Anthropocentrism and Its Discontents*, 103-11. For classical views of animals in general, see Sorabji, *Animal Minds*; and Newmyer, "Animals in Ancient Philosophy."

15 *TS*, 7, 3, 8: 220-5, esp. 220-1.

16 Ibid., 4, 1, 5: 14-23; 4, 2, 1: 101-10, esp. 109-10; 4, 3, 4: 217-22; 4, 4, 12: 348-52, esp. 348-9.

17 Ibid., 7, 2, 2: 121-9, esp. 124-5. On this see of course Lovejoy, *Great Chain of Being*.

18 *TS*, 6, 4, 16: 341-6, esp. 342-3.

19 Some of these examples were related to the connection between a belief in metempsychosis and vegetarianism. Mahmut alluded to the common belief in a Golden Age when animals were not eaten, and he related how the consumption of animal flesh arose from religious rites of animal slaughter. See ibid., 4, 1, 5: 14-23; and also 3, 2, 5: 150-2; 4, 2, 1: 101-10, esp. 109-10; 4, 4, 7: 328-32; 5, 2, 3: 87-90, esp. 89-90.

was a capital crime to shed the blood of any animal, the consumption of flesh was considered a sin, and how one even found hospitals for animals there. He acknowledged that not all Indians adhered to these precepts, yet for this and other reasons he expressed an ardent desire to travel to India.[20] The unfavorable comparison of western attitudes toward animals with those common in non-western cultures was also related to early modern primitivism in general. With its emphasis on the adverse effects of civilization, progress, and human reason, primitivism was naturally predisposed to criticism of scientific and philosophical novelties and to a favorable view of animals. It was therefore no surprise that a prominent primitivist such as Montaigne was also a theriophile. Mahmut displayed the same congruence of ideas.[21]

While primitivism was occasionally connected to a sincere appreciation of nature and animals, in most cases it treated them mainly as vehicles for human edification. This was also evident regarding the more literary aspects of early modern theriophily. Occasionally, however, more extreme theriophiles did accord animals attention in their own right. Mahmut did so when he wrote, in a vein critical of traditional biblical anthropocentrism:

> The Christians seem to have too proud an Opinion [of] themselves, and set a greater Value on Humane Nature than suits with Reason. They assert, that all Things were made for Man, and style him Lord of his Fellow-Creatures; as if God had given him an Absolute Dominion over the Rest of his Works, especially over the Animal-Generations; and that all the Birds of the Air, Beasts of the Earth, and Fish of the Sea, were Created only to serve his Appetite and other Necessities of Life.[22]

It was therefore no surprise that Mahmut was critical of the Cartesian beast-machine theory. He was generally an advocate of Descartes, the "admirable Genius" who was "renowned throughout the *Western World*, being esteemed the Best *Philosopher* that ever wrote of *Natural* Things." He professed a personal acquaintance with Descartes, claiming that they had often conversed with each other and that he would have believed him to be divinely inspired if not for his impious attitude toward Islam.[23] However,

20 Ibid., 6, 3, 5: 191-9; and also 4, 1, 5: 14-23; 4, 3, 4: 217-22; 5, 2, 3: 87-90, esp. 88-9.

21 For Mahmut's primitivism, see Tucker, "*Turkish Spy*," 88-91.

22 *TS*, 4, 3, 4: 217-22, at 217.

23 Ibid., 2, 1, 31: 109-12, esp. 111-12.

Mahmut was evidently censorious of the Cartesian view of animals. He explained and criticized it, claiming that the views of the Arabs and the ancient Greeks such as Pythagoras and Porphyry proved it to be erroneous. He could not believe that animal souls were mere matter, and he even implied that animals might have immortal souls, although he suspended his belief of this until he should "have the Happiness to see them there [in Paradise]."[24] After giving various theriophilic examples of animal sagacity he proceeded to claim:

> Can all these [animal] Actions proceed from meer Matter? In my Opinion, 'tis as easie to defend, That Humane Nature it self is but Matter so and so Modified, and, that all the Bustle Men keep in the World, is but the Effect of a better Composition of Body, the Result of a more perfect and refined Machine.[25]

Mahmut here anticipated Julien Offray de La Mettrie's later "man-machine" extrapolation of the beast-machine doctrine to explain also human actions. Yet like most subsequent eighteenth-century philosophers who rejected La Mattrie's extreme materialism, Mahmut seemed censorious of the concept of either human or animal material souls.[26]

Given all of the above, it is no surprise to find that Mahmut extended his theriophily from the theoretical to the practical-ethical realm. In fact, his explicit connection between a perception of the mental characteristics of animals and the existence of moral duties toward them was relatively more emphatic compared to most other examples of early modern theriophilic argumentation. He maintained this explicitly when writing that

> the Brute Animals ... have Souls as well as We ... [and] have Faculties and Affections conform[ing] to Ours. And therefore, it is little less Injustice to Kill and Eat them, because they cannot speak and converse with us, than it would be for a Canibal to murder and devour thee or me, because we understood not his Language nor he ours.[27]

24 Ibid., 2, 1, 26: 93-6.

25 Ibid., 96. Also see 4, 3, 4: 217-22.

26 For La Mettrie's moderate theriophilic views, see La Mettrie, *Machine Man*, vii-xxx, 9-14, 18-22, 26-7, 34-5, 38-9, 50, 52, 87-8, 97, 99, 100, 132, 139, 147, 150; and Hastings, *Man and Beast*, 44-5, 94-108.

27 *TS*, 4, 4, 3: 305-10, at 310.

And similarly:

> The Grand Root of the Common Injustice which Men are guilty of, in reference to the Beasts, and of the Intemperance with which they corrupt themselves, I perceive, is a false Principle which they have establish'd, denying the Capacity and Use of Reason to all Living Creatures but themselves.[28]

The fact that animals had similar but inferior mental characteristics compared to human beings was often utilized in the early modern era as an argument for disregarding their right to moral consideration, even if they were not regarded as machines. Mahmut, however, took the opposite view. His acknowledgement that human beings were superior to animals thus became essentially unimportant from an ethical perspective. The fact that animals were similar to human beings emphasized the need to treat them benignly, and the further fact that this similarity was incomplete did nothing to diminish this need. Nevertheless, he ineluctably weakened the overall force of his arguments for an ethical consideration of animals because he did not sufficiently apply them to the practical sphere in a detailed manner. In doing so he implicitly acknowledged that animals were not truly equal to human beings.

Many of Mahmut's remarks about the ethical treatment of animals, however, were outlined without making a specific connection to the issue of animal characteristics. For example, he related the story of a European Duke who savagely punished one of his poor subjects for unlawfully hunting a deer. Mahmut stated that had this been a punishment for the killing of the animal it would have been justifiable. But in reality the Duke himself hunted, so the punishment was unjust. Mahmut claimed that most of the great men of Europe were guilty of the "Murder" of animals in the practice of hunting, and he denounced them as those "whose Tables are no other than the *Altars* of *Gluttony*, smoaking with Flesh and Blood, whilst *Hecatombs* of Animals are there sacrific'd to Voracious Appetites, the *Idols* of these *Western* People."[29] Other contemporaries expressed similar sentiments. Earlier in the seventeenth century, Margaret Cavendish, the Duchess of Newcastle, had written of those who make "their Stomacks Graves, which full they fill / With Murther'd Bodies, which in sport they kill."[30] And in the following century, Alexander Pope, one of the most committed theriophiles of his era,

28 Ibid., 7, 3, 8: 220-5, at 220. The "corrupting intemperance" was the consumption of meat.

29 Ibid., 5, 2, 2: 82-6; 5, 2, 3: 87-90, 88 for the quotation.

30 Cavendish, "The Hunting of the Hare," in *Poems*, 158-62. See also Nelson, "'Worms in the Dull Earth of Ignorance'."

wrote that he knew "nothing more shocking, or horrid, than the prospect of ... kitchens covered with blood, and filled with the cries of creatures expiring in tortures."[31] Mahmut reiterated his vegetarian proclivity when he stated:

> But our Reason in this Point [the praise of vegetarianism], ought to take its Rise from the Fundamental Law of Nature, the Original Justice of the World, which teaches us, Not to do that to another, which we wou'd not have another do to us. Now, since 'tis evident, That no Man wou'd willingly become the Food of Beasts; therefore, by the same Rule, he ought not to prey on them.[32]

If eating animals was admissible, then people might as well eat slaves and war prisoners, over whose life and death one also had a right, according to the law of nations.[33] A vegetarian diet was therefore not only morally imperative, but also ethically beneficial in a general sense, since gluttony had adverse moral consequences, while temperance and abstinence were important for virtue as well as purity of spirit. Vegetarianism was healthy both to body and mind and assisted the perfection of human nature.[34]

Yet, despite his moral advocacy of such a diet, Mahmut was unable to sustain it in practice, and not just because of his need, as a spy, to conceal his Muslim identity by drinking wine and eating swine's flesh.[35] This self-conscious limited commitment to practice what one preached was a common trait of early modern theriophily. Even Montaigne had exhibited it in his *On Cruelty*, where alongside criticism of harshness toward animals and a plea for treating them benignly he acknowledged that he was nevertheless not absolutely committed to his own precepts. Together with a clear detestation of animal suffering, he also expressed an ostensibly self-ironic predilection for hunting.[36] And in the eighteenth century, Rousseau, as we will see later

31 For this quotation, and Pope's general theriophilic philosophy, see Pope, *Works*, 10: 515-21; and also Ault, "Pope and his Dogs"; Nicolson and Rousseau, *"This Long Disease, My Life"*, 93-109; Thomas, *Man and the Natural World*, 110, 119, 140, 149, 183-4, 289; Harwood, *Love for Animals*, 139, 174-85, 297, 301-2.

32 *TS*, 7, 2, 7: 145-54, at 150-1. In this letter, however, Mahmut implied that eating locusts, crabs, and crawfish, was morally permissible.

33 Ibid., 7, 3, 8: 220-5, esp. 224. This type of reasoning implied the constant theme in western philosophy connecting attitudes toward animals with attitudes toward human beings. This was not necessarily a theriophilic argument, since it could also imply that animals were important only as a means to human ends.

34 Ibid., 3, 2, 15: 66-70; 6, 3, 5: 191-9, esp. 193-4; 7, 2, 2: 125-9; 7, 2, 7: 145-54. For examples of vegetarianism, see also 4, 1, 5: 14-23.

35 Ibid., 2, 3, 35: 351-5.

36 Montaigne, "On Cruelty," in *Complete Essays*, 472-88.

on, exhibited this same ironic self-limitation to an even greater degree. *The Turkish Spy* in this, as in a more general sense, exemplified the persistence of the classically-minded early modern theriophilic tradition into the long eighteenth century. Like Montaigne and Rousseau, Mahmut was aware of his own moral infirmity and was candid about this weakness. Repeatedly, as he praised vegetarianism, he expressed his regret at not being able to maintain it. He wished that he might practice what he so admired "and not be self-condemn'd for living contrary to" his "Knowledge. For God neither loves a double Tongue or Heart."[37] He referred to himself as one who had "often attempted to live in *Abstinence*; but, by the Force of a Voracious Appetite, suffered" himself "to be carri'd back to" his "old Intemperance."[38]

Mahmut's debility of will to practice his moral convictions mitigated the overall efficacy of his theriophilic arguments. It was also accompanied by occasional derogatory references to animals, which, while not in themselves denying them moral consideration, exemplified a distinctly un-theriophilic viewpoint. For example, while mentioning that the Christians believed that the Divinity resided in their temples, he added that nevertheless "they suffer Dogs to Profane them with their vilest Excrements"; and that they adorned their churches and altars with invaluable treasures, "and yet, after all, they must become the Receptacles of the Dung of Sordid Animals."[39] He also referred to swine as "Filthy Animals," although this may have been due to his Islamic faith.[40] No such explanation, however, could be given for his praise of the father of one of his correspondents for being a great hunter of fierce animals, "Savages" and "Ravenous Creatures" such as tigers, panthers, and especially lions.[41] Elsewhere he related having almost swallowed a spider that had drowned in his wine, a creature to which he had "an *Invincible Aversion* and *Abhorrence*." His explanation for this repugnance was that if the theory of metempsychosis was true, then perhaps he had been a fly in a previous life, and simply remembered his antipathy for his former enemy.[42] Not too much should be made of these infrequent remarks, however. For example, at one point he had mentioned, as noted above, that he derived pleasure from observing spiders. This would tend to support the thesis of multiple authorship of *The Turkish Spy*, in which case not all the authors

37 *TS*, 7, 2, 7: 145-54, esp. 154.

38 Ibid., 4, 2, 1: 101-10, at 110. See also 3, 2, 15: 66-70; 4, 3, 4: 217-22, esp. 218; 7, 2, 2: 121-9; 7, 3, 8: 220-25.

39 Ibid., 2, 1, 23: 74-7, at 76.

40 Ibid., 5, 2, 2: 82-6, at 86.

41 Ibid., 3, 2, 5: 150-2.

42 Ibid., 4, 4, 12: 348-52.

would necessarily have shared the same theriophilic beliefs, even though those preponderated in the text.

The physical and mainly spiritual benefits of a vegetarian diet were clearly salutary. However, Mahmut also raised a religious advantage for treating animals considerately, since God "will assuredly grant us a perpetual Tranquility, if we abstain from injuring our *Fellow-Animals*"[43] and, as he admonished one of his relatives, "serve *God* after the Manner of thy *Forefathers*: love thy Friends, pardon thy Enemies, be Just to all Men, and do no Injury to any *Beast*. If thou observest this Rule, thou may'st defy the *Devil*; for thy *Soul* is in Safe Custody."[44]

A direct influence of *The Turkish Spy* on specific cases of eighteenth-century reasoning about animals is not easily traceable. Edward Gibbon, whose attitude toward animals we will examine in greater detail in a later chapter, was not insensitive to animal suffering and in his early *Essai sur l'étude de la littérature* expressed abhorrence at the killing of animals in the ancient Roman spectacles.[45] However, in *The History of the Decline and Fall of the Roman Empire* he regarded the human dominion over the animals, and specifically the domestication of animals, as culturally beneficent.[46] Gibbon owned a copy of *The Turkish Spy*, but to what extent it influenced his view of animals remains difficult to ascertain, though it does not seem to have been considerable.[47] A more direct link of this sort was exhibited by Adam Smith, whose innovative view of animals will also be examined in detail in a later chapter. In his *Theory of Moral Sentiments* he related the following:

> Those animals ... that have been remarkably serviceable to their masters, become the object of a very lively gratitude. We are shocked at the brutality of that officer, mentioned in the Turkish Spy, who stabbed the horse that had carried him across an arm of the sea, lest that animal should afterwards distinguish some other person by a similar adventure.[48]

The tone of this passage suggested that Smith took for granted that his readers were familiar with *The Turkish Spy* and perhaps also shared its sentiment for animals. This in itself is further testimony of the eighteenth-century popularity of the work in and outside of Britain. It directly influenced famous

43 Ibid., 4, 4, 3: 305-10, at 310.
44 Ibid., 5, 4, 6: 301-6, at 305.
45 Gibbon, *Essai*, chapter 41.
46 Gibbon, *History of the Decline and Fall*, 3: 154; also 1: 1026-7.
47 Keynes, *Library*, 190.
48 Smith, *Theory of Moral Sentiments*, 94-6.

works such as Montesquieu's *Lettres persanes* and Goldsmith's *Citizen of the World*. If the former of these contained no significant mention of animals,[49] in letter 15 of the latter work Goldsmith condemned the hypocrisy of the English, who professed sensitivity to animals and yet continued to eat them. "Strange contrariety of conduct! they pity, and they eat the objects of their compassion! ... Man ... was born to govern the brute creation, but he has become their tyrant." Goldsmith implied the possible existence of metempsychosis and also wrote:

> Hail, O ye simple, honest brahmins of the East; ye inoffensive friends of all that were born to happiness as well as you; you never sought a short-lived pleasure from the miseries of other creatures! You never studied the tormenting arts of ingenious refinement; you never surfeited upon a guilty meal! How much more purified and refined are all your sensations than ours! you distinguish every element with the utmost precision; a stream untasted before is new luxury, a change of air is a new banquet, too refined for Western imaginations to conceive.[50]

All this was redolent of *The Turkish Spy*'s influence, specifically the attention given to the Indian treatment of animals and the moral argumentation for vegetarianism. Samuel Johnson, who undoubtedly was familiar with *The Turkish Spy*, may also have been similarly influenced, given his mild theriophilic beliefs.[51] He abhorred vivisection, and James Boswell related the particular care he lavished on his cat Hodge. Nevertheless, he did not deny human beings' basic right to kill animals for food, even if only without unnecessary cruelty. He also seems to have denied animals reason, explaining their actions as guided by instinct.[52] Gibbon, Smith, Goldsmith, and Johnson testify to the popularity of *The Turkish Spy* in Enlightenment Britain and to the likely influence of its theriophilic views. Yet it is significant that they all avoided the more extreme aspects of this theriophily. In eighteenth-century Britain it was popular to express concern for animal suffering, but only to a moderate extent. This was a respectable attitude only so long as it did not offend social sensibilities by exhibiting any kind of parity between humans and animals.

49 But see the brief allusions to vegetarianism and metempsychosis in letter 46, and to animal experimentation and the beast-machine theory in letter 145, in Montesquieu, *Persian Letters*, 101-2 and 265-8 respectively.

50 Goldsmith, *Works*, 191-2.

51 For this familiarity, see McBurney, "Authorship," 922-3.

52 "The Idler" no. 17, in Johnson, *Idler and The Adventurer*, 53-6; Boswell, *Boswell's Life of Samuel Johnson*, 154-5, 197-8, 202-3, 434-5, 450-1, 486.

A common interpretation among historians has been that a certain improvement in attitudes toward animals occurred toward the end of the eighteenth century. It was gradual and not revolutionary, yet not insignificant. This is true as far as it goes, although, as we have already noted, it was also a manifestation of an urban civilization increasingly removed from nature and in control of it. Historians are of course also aware that abuse of animals did not cease. Some of them have discerned a shift from a more scientific to a more purely ethical type of discussion regarding animals, which prepared the ground for what is usually considered as the relatively more humane treatment of animals from the late eighteenth century onwards.[53] However, it is a consistent theme of the present book that while a significant shift did occur in considerations of animals toward the late eighteenth century, it was not in a more ethical direction. As we have seen, an ethical concern with the treatment of animals was a constant theme in early modern intellectual culture, in itself much indebted to classical sources. The originality of the eighteenth and nineteenth centuries in this respect has been consistently overblown. The shift that did occur was in fact in a scientific direction, though a new one: not the philosophical rationalistic discussion of animal mental characteristics which had been a hallmark of both the theriophilically- and the Cartesian-inclined positions, both of which were effectively passé by the late eighteenth century, nor the physiological-anatomical investigation of animal bodies, which had originated in the Renaissance and gradually intensified till modern times, but, rather, a shift toward a more modern historiographical, and eventually economic, consideration of the place of animals in human civilization. This change will be the subject of most of the following chapters, but it is interesting at this point to already note that both Gibbon and Smith were significant protagonists in its development.

As for the more traditional interpretation noted above, particular attention has repeatedly been given by historians to Jeremy Bentham as an exemplar of the purported rise of modern sensitivity to animal suffering. Bentham had claimed that what was important about defining the treatment of animals was not whether they could reason or talk but whether they could suffer. Yet what is not usually noted is that Bentham, while opposing causing unnecessary pain to animals, did not in principle oppose killing them for food, or even practicing animal experimentation, as long as it was

53 E.g. Thomas, *Man and the Natural World*, passim; Harwood, *Love for Animals*, 64, 74-5, 126, and passim; Hastings, *Man and Beast*, 16; Bradie, "Moral Status"; Floridi, "Scepticism"; and Maehle, "Cruelty and Kindness."

beneficial for humanity.[54] Nowhere did he claim that animal feelings were on a par with human emotions, and in fact he never contested their essential inferiority compared to human beings. He was therefore not that different in his view of animals from most previous early modern theriophiles. Like Montaigne, Mahmut, and Rousseau, his sensitivity to animal suffering was severely diluted in the transition from theory to practice. It should also be noted that a specific translation of the notion of moral sensitivity to animals to an actual legal concept of animal rights has also been located by some historians in eighteenth-century theriophilic argumentation, for example in the thought of Francis Hutcheson. In fact, however, it can be located much earlier. The 1641 Massachusetts Body of Liberties, which was in effect a Bill of Rights, alongside of course mainly consideration of civil rights for humans, also stipulated that "No Man shall exercise any Tirrany or Crueltie towards any brute Creature which are kept for man's use."[55] But the point which needs to be made is that a notion of the right animals had to moral consideration, more than its specific formulation in philosophical or legal terms, was already prevalent by the late seventeenth century.

Theriophiles throughout the early modern era had been as preoccupied as those of the late eighteenth century with animal suffering, even if they had not always made a point of specifically emphasizing emotions, rather than intellectual abilities, as an argument for according animals ethical consideration. Mahmut expressed a concern for animals both in connection to their mental characteristics but also, anticipating Bentham, with due recognition that the inferiority of these characteristics compared to those of human beings was not in itself an impediment to this moral concern. The claim that animals deserved moral consideration irrespective of their mental abilities was an unconventional moral precept, however, in discord with the traditional notion of human uniqueness. Most early modern theriophiles, including the fictional figure of Mahmut, exemplifying the plural authorship of *The Turkish Spy*, found adherence with such a view difficult to maintain. The issue of animal characteristics was usually considered a basic premise of ethical debates about animals, and the mental and spiritual superiority of human beings over them was rarely contested, even among theriophiles. Montaigne, Mahmut, Rousseau, and Bentham all considered animals as worthy of moral consideration, but at the same time they all refrained from equating this consideration with that accorded to human beings. There was no significant change in this respect throughout the early modern era.

54 Bentham, *Introduction*, 44, 282-3; Bentham, "Letter"; and also Bradie, "Moral Status," 42.
55 Commager, *Empire of Reason*, 218. On Hutcheson, see chapter 6 below.

4 Rousseau and Animals

The tradition of theriophily, broadly defined, persisted during the eighteenth century. One of its main proponents was Jean-Jacques Rousseau. It should be noted at the outset that Rousseau was far from a typical Enlightenment figure. His criticism of the ills of advanced civilization was more in line with primitivistic philosophy, and it is therefore no surprise to see that he expressed theriophilic opinions, since primitivism and theriophily were often conjoined. It is also no surprise that his theriophily was, like that of his predecessors, limited in the transition from theory to practice. Rousseau's works contain many references to animals, most of them in passing, but others much more substantive and touching on central aspects of his thought. This chapter will not present a detailed survey of these sources, but will highlight only what seems important for a better understanding of key elements of Rousseau's moral attitude toward animals, and what these can teach us both about his general philosophical outlook and about general trends in the Enlightenment's consideration of this topic.[1]

The Victorian humanitarian Henry Salt claimed at the end of the nineteenth century that "it was not until the eighteenth century, the age of enlightenment and 'sensibility', of which Voltaire and Rousseau were the spokesmen, that the rights of animals obtained more deliberate recognition."[2] Voltaire had indeed attacked the Cartesian beast-machine theory, and in general all the philosophical debates about animal mental and spiritual characteristics, singling out the cruelties of vivisection, which the view of animals as non-sentient beings helped popularize.[3] Rousseau repeatedly addressed the issue of animals even more than Voltaire, and in the *Discourse on Inequality*, for example, claimed that human moral conduct was directed primarily by commiseration, not rationality. Animals, although denied human enlightenment, were nevertheless similar to human beings in their sensitive capacity, and human beings had some type of duties toward them. "It seems, in effect, that if I am obliged not to do any harm to anyone similar to me, this is less because he is a reasonable creature than because he is a sensitive one; a quality being in common among the

1 For a different interpretation of Rousseau's attitude toward animals, see Guichet, "Animality." The most conclusive survey remains Hastings, *Man and Beast*, 113-21, 259-61. The vegetarian issue is best explored in Boonin-Vail, "Vegetarian Savage." Also see Steiner, *Anthropocentrism and Its Discontents*, 177-80.

2 Salt, *Animals' Rights*, 3.

3 Voltaire, *Dictionnaire philosophique*, 411-15 (s.v. "Bêtes"); Voltaire, "Dialogue."

animals and the man, it should at least give the one the right not to be maltreated unnecessarily by the other."[4] However, we have already seen that such claims were expressed long before, in various ways, by Montaigne and others. Salt's simple evaluation of Rousseau's view of animals therefore misses many historical intricacies and qualifications, and the same can be said of most subsequent discussions of his view of animals.

In the *Discourse on Inequality* Rousseau claimed that animals in general had a certain degree of ability to think, yet human beings were the only animals capable of contributing to their operations by their capacity as free agents. Human beings were also uniquely capable of self-perfectibility, both as individuals and as a species. Yet this was also the source of human misery, since these abilities drew them from their initial tranquil state and resulted in both the human virtues and vices.[5] This was in the vein of Renaissance theriophily, which had made a point of denigrating human vices and weaknesses by comparing them to animals. In the *Discourse on Inequality*, of course, Rousseau was primarily interested in clarifying his primitivistic criticism of the ills of modern urban civilization and in his advocacy of the more innocent state of primitive humanity. The details of his views on this issue, and the exact extent of his primitivism, remain to this day one of the more contentious issues in Rousseau scholarship, and we cannot here go into too many details.[6] Arthur Lovejoy observed long ago that Rousseau extolled not the absolute state of nature but, rather, an intermediate state in between pure primitivism and advanced civilization.[7] This moderation, not always evident in Rousseau's intellectual outlook, was also applicable to his consideration of animals.

Rousseau was typically not interested in the more cognitive aspects of the debate on animals; the ethical aspect interested him much more. In *Les Confessions* he related how as a child he never delighted in torturing animals.[8] He had even thrown stones at animals he saw tormenting other

4 *Discours sur l'origine et les fondemens de l'inégalité parmi les hommes*, in Rousseau, *Oeuvres complètes* [hereafter cited as *OCR*], 3: 126: "Il semble, en effet, que si je suis obligé de ne faire aucun mal à mon semblable, c'est moins parce qu'il est un être raisonnable que parce qu'il est un être sensible; qualité qui étant commune à la bête et à l'homme, doit au moins donner à l'une le droit de n'être point maltraitée inutilement par l'autre." See also Thomas, *Man and the Natural World*, 175-6.

5 *OCR*, 3: 141-2.

6 For a general evaluation, see Kelly, "Rousseau's 'Peut-Etre'," which ends with an interesting remark on Rousseau's seeming view of human beings as animals.

7 Lovejoy, "Supposed Primitivism."

8 *OCR*, 1: 10.

weaker animals,[9] an interesting extrapolation of his sense of justice from the human to the animal world – if humans owed justice to animals, then the latter also owed it to each other! Like many upper-class eighteenth-century people, he owned pets.[10] Moreover, he delighted in taming animals, for example pigeons, inspiring them with a confidence which he never abused and claiming that "All animals distrust man and justifiably so; but once they are sure he means them no harm, their confidence becomes so great that one must be more than a barbarian to abuse it."[11]

These observations had a specific moral consequence evident in Rousseau's advocacy of vegetarianism. In the notes to the *Discourse on Inequality* he claimed that human beings were anatomically vegetarians. Since vegetarians, in contrast with carnivores, lived peaceably one with the other, this meant that had human beings remained vegetarians, they would have had an easier time subsisting in the state of nature, and they would have had less need to leave it.[12] In *Emile ou de l'éducation* Rousseau stated that people should eat less meat and more vegetables in order to maintain better health.[13] Children were indifferent to meat and preferred eating vegetables. Furthermore, carnivorous nations such as England were crueler than vegetarian nations.[14] Commenting on the rise of the feeling of pity in Emile, Rousseau noted the need to commiserate with the suffering of others, including animals.[15] People pitied the suffering of others only to the extent that they perceived that suffering. Therefore, just as rich people did not sufficiently pity poor people, so human beings did not sufficiently pity animals, because they thought that animals were unaware they were going to be slaughtered or were unable to think of their past or future suffering.[16]

Rousseau's moral observations on animals were broader in scope than just vegetarianism, however, and touched on his whole ethical outlook. This emphasized commiseration, or as he and other contemporary philosophers such as Hume or Adam Smith often called it, "sympathy," with the suffering of others. Nowhere was this more evident than in his best-selling epistolary

9 Ibid., 1: 20.

10 Ibid., 1: 521, 556-7.

11 Ibid., 1: 240: "Tous les animaux se défient de l'homme et n'ont pas tort; mais sont-ils sûrs une fois qu'il ne leur veut pas nuire, leur confiance devient si grande qu'il faut être plus que barbare pour en abuser." See also ibid., 233-4.

12 Ibid., 3: 198-9.

13 Ibid., 4: 274-6.

14 Ibid., 4: 411-14.

15 Ibid., 4: 505-6.

16 Ibid., 4: 508-9.

novel *Julie, or the New Heloise*. In the novel, St. Preux, while recounting the virtuous life of Julie and her husband, M. de Wolmar, related how, to his surprise, they regarded themselves as the guests of the birds in their garden, not the other way around. They made an effort to keep the enemies of the birds, and especially children, away from these animals. The birds stayed in the garden of their own free will, in contrast to the fish, who found there a safe haven from the frying pan at the price, undoubtedly worth it, of living confined in a basin.[17] St. Preux also related how while sailing with Julie and some servants on Lake Geneva he wanted to shoot some birds, only to be shamed by Julie for his desire to do harm to these animals. While this was a fishing expedition, Julie preferred letting almost all the fish back into the water, despite the wishes of St. Preux and the servants. She regarded even fish as suffering animals and wanted to enjoy their having evaded danger.[18]

Elsewhere in the book Rousseau implied that hunting should be criticized, and he viewed the suffering of horses as among the evils (which included primarily human suffering) which one encountered in badly governed countries. So far so good. However, in the same passage Rousseau included a short depiction, not particularly critical, of how St. Preux and Julie's father went hunting birds together.[19] Here for the first time we begin to encounter the limits to Rousseau's theriophily. By depicting Julie as more critical of hunting than the men, Rousseau seemed to confess that he was not categorically against hunting. There was a gender element to this. One might claim that the sensitivity to animal suffering was here a concomitant of the "weak sex," thus to be tolerated sympathetically. It should also be remembered that Rousseau, though not a misogynist, was not a feminist, even by eighteenth-century standards. At the same time Julie, even more than Emile, was the figure in Rousseau's writings who more than any other represented a purely virtuous existence. In any event, a theriophilic sensitivity was customarily part of a virtuous ethics, according to Rousseau.

Nevertheless, he did not live up to his own ethical admonitions. It was clear from *Les Confessions* that he was not a vegetarian, and although not a hunter he was not above "making war" against the dormice which he thought ate M. d'Epinay's garden fruit; he destroyed many of them, but to no avail, since the gardener turned out to be the culprit.[20] Like Montaigne, Mahmut, and Bentham, his theriophily was diluted in the transition from

17 Ibid., 2: 475-8.

18 Ibid., 2: 514-15.

19 Ibid., 2: 603-6. On this novel and animals, see also Robbins, *Elephant Slaves*, 198-9.

20 *OCR*, 1: 433, 507, 543.

theory to practice. He even recognized the need for dissecting animals in order to study them (he did not specifically mention experimenting upon live animals, but this seemed to be implied). However, he viewed anatomical theatres as frightful places, claiming that "It is not there, upon my word, that J. J. will go in search of his amusements."[21] Nowhere were the limits to his theriophily more clear than in *Emile*, his blueprint for the correct way to educate the ideally virtuous and unspoiled human being. One of the great educational difficulties to be overcome was that of rising adolescent sexual awareness. Rousseau was interested in presenting the theme of love to Emile in a gradual manner, stifling the stronger amorous passions before they did early damage. Therefore, as a distraction, he recommended that Emile practice hunting, despite its cruelty, claiming that "I do not want that all Emile's youth be passed in killing animals and I do not even pretend to justify completely this ferocious passion; it suffices for me that it serve enough to suspend a very dangerous passion, in order for him to listen to me calmly as I talk of it, and give me the time to paint it without exciting it."[22] Here as elsewhere in his writings, Rousseau encountered the inflexibility of reality, which made the attainment of perfect virtue so difficult. However, in the Sisyphean attempt to overcome the obstructions on the way to moral transparency, to use Jean Starobinski's terminology, a price had to be paid.[23] This price was the relinquishing of perfection, and when human and animal interests collided there was no doubt that the latter paid this price.

There was, however, a deeper significance to all this. For Rousseau, the ethical dilemmas regarding animals were only one part of his general *Weltanschauung*, of his continual striving for a moral existence, and the attempt to make sense of it all, in the face of an unremittingly evil-ridden world. His views on animals were thus ultimately, like those of most other early modern theriophiles, part of a more general anthropocentrically-driven primitivistic critique of human corruption. The more sentimental expressions of this theriophily, to which *Julie* gave such an important and early utterance, would have to wait till the nineteenth century for more common appearances. But even more important is the fact that in contemplating animals, Rousseau was metonymically confronting his most deep-seated personal philosophical

21 *Les rêveries du promeneur solitaire*, in *OCR*, 1: 1068: "Ce n'est pas là, sur ma parole, que J. J. ira chercher ses amusemens."

22 Ibid., 4: 644-5: "Je ne veux pas que toute la jeunesse d'Emile se passe à tuer des bêtes et je ne prétends pas même justifier en tout cette feroce passion; il me suffit qu'elle serve assés à suspendre une passion plus dangereuse pour me faire écouter de sang-froid parlant d'elle, et me donner le tems de la peindre sans l'exciter."

23 Starobinski, *Jean-Jacques Rousseau*.

difficulties. A striking expression of this was given in his famous "Letter on Optimism" (or "Letter on Providence," as it is occasionally called) to Voltaire, one of the most important and revealing of his writings. In this letter Rousseau evinced his belief in immortality and in some type of just God, even if one who had no particular interest in human activity. It was people, not God, who were responsible for the world's ills. The individual human being was insignificant from a universal and divine perspective yet could still contribute to the common good, even by personal difficulties.[24]

> That the human corpse nourishes worms, wolves, or plants, this is not, I confess, a compensation for that person's death; but if, in the system of this universe, it is necessary for the conservation of the human race that there would be a circulation of substance among the humans, the animals and the vegetables, then the particular evil suffered by the individual contributes to the general good. I die, I am eaten by worms, but my children, my brothers will live as I have lived, my corpse fattens the earth whose productions they will eat, and I do, according to the natural order and for all men, that which Codrus, Marcus Curtius, the Decii, the Phileni [all ancient heroes who sacrificed themselves for their countries], and a thousand others, voluntarily did, for a small part of humanity.[25]

Here, with these Hamlet-like remarks, was a key to implicitly overcoming the obstacle of eating animals. Death was inevitable, and just like the death of a human being was beneficial in the grand order of things, so, *a fortiori*, was the death of an animal. Yet there was no Panglossian conclusion here. As is well known, this letter was an important step in the philosophical and personal worsening in the relationship between Voltaire and Rousseau, with the latter clinging to a view of a future compensatory immortality in the face of the former's criticism of the idea of a just providence interested in

24 See the letter to Voltaire from August 18, 1756, in Rousseau, *Correspondance générale*, 2: 303-24, at 315.

25 Ibid., 315-16: "Que le cadavre d'un homme nourrisse des vers, des loups, ou des plantes, ce n'est pas, je l'avoue, un dédommagement de la mort de cet homme; mais si, dans le système de cet univers, il est nécessaire à la conservation du genre humain qu'il y ait une circulation de substance entre les hommes, les animaux et les végétaux, alors le mal particulier d'un individu contribue au bien général. Je meurs, je suis mangé des vers, mais mes enfans, mes frères vivront comme j'ai vécu, mon cadavre engraisse la terre dont ils mangeront les productions, et je fais, par l'ordre de la nature et pour tous les hommes, ce que firent volontairement Codrus, Curtius, les Décies, les Philènes, et mille autres, pour une petite partie des hommes." For the importance of this letter, see Leigh, "Rousseau's Letter"; Leigh, "From the *Inégalité* to *Candide*"; and Havens, "Voltaire, Rousseau."

human affairs. As with other issues, there was more similarity than difference between the views of the two savants, yet it was the differences which predominated in their correspondence, and probably also in the estimation of their contemporaries. Thus both of them demonstrated a similar ethical view of animals, but while Voltaire had no problem articulating it in a clear and straightforward manner, advocating sensitivity to animals while retaining a clear sense of human superiority, Rousseau typically tortured himself with every imaginable philosophical intricacy related to animals, and to every other ethical topic for that matter. In doing so, however, he left many important traces for scholars interested in the finer points of these issues.

Rousseau was far from a typical or representative figure of Enlightenment thought, yet we should constantly remember just how influential he was during the late Enlightenment and early Romanticism. There was probably no highly educated European during this era who was not familiar with most, or even all, of the works mentioned above, the letter to Voltaire excepted. His ideas about animals in particular were, compared with other topics he addressed, quite in tune with earlier intellectual developments. On a broader level, in almost all early modern discussions of the human interaction with the natural world there was a preponderant underlying assumption regarding the human anthropocentric claim of entitlement, emphasizing humanity's self-conception of itself as the apex of the natural world destined to control and utilize it for its own purposes. While initially this was a religiously motivated view, it was later assimilated within the more secular-rational outlook of the Scientific Revolution and subsequently of the Enlightenment.[26] The history of this anthropocentric outlook is almost surprisingly continuous and uninterrupted in western culture, and it underlies all but the most rare instances of ethical considerations of animals, even those emphatically theriophilic.[27] Rousseau's sentimental expressions of sensitivity to animals were both a summation of the early modern theriophilic tradition and an important influence on the subsequent romantic susceptibility to this outlook. Yet in both senses the clear limits to his moral sensitivity to animals suggest that there was much less of an increase in this sensitivity during the long eighteenth century than is often assumed. Rousseau embodied the high point of early modern theriophilic sensitivity to animals. As we will see below, and as *The Turkish Spy* had already demonstrated, it was possible to express an even more impassioned susceptibility to animals, or one, as in Bentham's case, more clearly enunciated, yet these were all ultimately

26 Harrison, *Fall of Man*.

27 Steiner, *Anthropocentrism and Its Discontents*; Wolloch, *Subjugated Animals*.

expressions of the same moral position going back at least to Montaigne. There is no doubt, as the following two chapters will demonstrate, that this was a broadly prevalent approach and in this respect deserves scholarly attention. Yet it should be remembered that ubiquity, its significance notwithstanding, does not mean originality. For that we will need to examine different aspects of Enlightenment thought.

5 William Smellie and the Enlightenment Critique of Anthropocentrism

The Enlightenment outlook inherited and expanded the Scientific Revolution's legacy of extolling the material mastery of nature. Ultimately, this legacy developed into the modern anthropocentric ethics of utilizing natural resources. Recent scholarship has subjected the generalizing term "Enlightenment" to various criticisms, differentiating between various national outlooks or between radical and moderate Enlightenments. There is, however, a general sense in which the term remains applicable to the eighteenth-century broad battle with religious superstition and political despotism. Another important general element of the Enlightenment *Weltanschauung*, not always sufficiently emphasized, was the advocacy of the need to command the forces of nature and harness them for human benefit and progress. During the eighteenth century the translation of traditional anthropocentric cosmology into scientifically-based, practical material benefits, became an unprecedented cultural endeavor. At the same time, as we will see, new forms of consideration of both the natural environment and animals began subsuming older ones, all the while retaining a basic anthropocentric premise.

Since the rise of the modern environmental movement, particularly from the 1960s, philosophers have consistently been interested in clarifying the logical and ethical dimensions of the concept of anthropocentrism.[1] The term "anthropocentrism" is a modern one, yet the concept, the notion that humanity stood at the apex of the material creation and that the creation was meant for human use, and conversely the critique of this outlook, have been a part of the western tradition since antiquity. We will therefore use this term here to refer to equivalent ideas before and during the long eighteenth century. The ancient and medieval Judeo-Christian cosmology based its anthropocentric outlook on divine sanction. The Scientific Revolution transformed, rather than replaced, this tradition. The result was the modern scientific anthropocentric ethics of technological mastery of natural resources. In our own time this has evolved, at least in some respects, into an environmental crisis, a development which would have truly surprised eighteenth-century intellectuals.

1 Passmore, *Man's Responsibility*, 3-40; Bayertz, "Nature of Morality."

This chapter will examine this topic by concentrating on the views of the semi-forgotten figure of William Smellie, an important and highly learned printer, journalist, and editor in eighteenth-century Edinburgh. Despite his many activities in these fields and the financial pressures of providing for a large family, Smellie, who was influential as a disseminator of learning in general, also found time to work on his own intellectual and scholarly projects. These included outlining one of the most original and detailed eighteenth-century critiques of anthropocentrism. Few intellectuals before modern times devoted attention to this topic more than Smellie. However, his importance and originality in this respect have received little attention from scholars, despite some recent growing interest in his activities and intellectual accomplishments. Smellie has been cited as claiming that the aim of human industry was to diminish noxious animals and augment useful vegetables.[2] Yet in grappling with the ethical dimensions of anthropocentrism, Smellie's ideas were in fact much more complicated. In so doing, as we shall see, Smellie was even more consistent and systematic than other more famous critics of anthropocentrism, such as Montaigne and Rousseau. Montaigne's theriophily was mainly an offshoot of his primitivism, and both his and Rousseau's views of animals were never systematically elaborated, let alone, in the case of Rousseau, in relation to the predominant contemporaneous emphasis on the rational scientific investigation of the natural world. This was precisely what Smellie offered. The tendency to concentrate on canonical figures does not always give a thorough understanding of the history of ideas.

Smellie is primarily remembered as an important figure in the printing world of late eighteenth-century Edinburgh, as editor of the first edition of the *Encyclopaedia Britannica*, as the most important English translator of the Comte de Buffon's *Histoire Naturelle*, and as the author of *The Philosophy of Natural History*, his own original venture into scientific work.[3] However, most studies about him have centered mainly on his social role in the world of the late Scottish Enlightenment, and almost no attention has been devoted to his own philosophical ideas. This modern neglect, and Smellie's failed struggle to become a prominent intellectual figure in Enlightenment Edinburgh, may in part be due to his being an amateur scientist, at a time

2 Worster, *Nature's Economy*, 51.

3 The most important source for Smellie's life and work remains Kerr, *Memoirs*; Sher's introduction to this modern reprint is also informative. For his activities as a major printer, see Brown, "William Smellie and Natural History." For his editorship of the *Encyclopaedia Britannica*, see Kafker, "William Smellie's Edition." For *The Philosophy of Natural History*, see Wood, "Introduction." For his translation of Buffon, see Garrett, "Introduction."

when professional scientific specialization was increasingly becoming the academic norm.[4] Furthermore, his low social status differentiated him from most of his Scottish Enlightenment colleagues. His financial difficulties and dependency on patrons such as Lord Kames created an indignity which gradually turned him into a political Whig, and he was therefore pursued by controversy, particularly in his later years. His social status perhaps contributed to his support of the democratization of learning.[5] It is more than likely that this life-long sense of being an outsider helped form his emphatic Whig sensibilities and also influenced his outlook on the natural creation. A general sense of injustice in human society probably also led to sensitivity to human injustice toward animals. In this respect, he was similar to Rousseau, although this similarity should not be pressed too far. Ethical anti-anthropocentrism was in any event not new in itself. What was original was the detailed argumentation which Smellie brought to this topic.

This originality was evident particularly in his essay *Whether Are All Animate and Inanimate Bodies Made for the Immediate Use and Conveniency of Mankind; or, is That Only a Secondary End of Their Existence?*, an early work written when he was twenty years old, and read before the Newtonian Society in Edinburgh in 1762.[6] Smellie at this early stage already set forth his main anti-anthropocentric arguments. We shall begin with these innovative, youthful ideas and then proceed to examine his similar mature views, particularly in comparison with the more famous opinions of Buffon. When Smellie began developing his interpretation of the human attitude toward nature, he was probably not yet intimately familiar with Buffon's works. The majority of these, in any event, had not yet been published. Buffon, as we will see, outlined a particularly emphatic anthropocentric outlook. However, it is a measure of Smellie's independent philosophical originality that, despite this influence, he adhered to his anti-anthropocentric ethics throughout his life. Already in an early letter from 1761 he disagreed with the view that human beings were the only social creatures, claiming, on the contrary, that animals were more social than humans.

4 Emerson, "Scottish Enlightenment," 61; Shapin, "Property, Patronage," 12-22.

5 Brown, "William Smellie 1740-95."

6 Smellie, "Whether Are All Animate and Inanimate Bodies Made for the Immediate Use and Conveniency of Mankind; or, is That Only a Secondary End of Their Existence?" in *Literary and Characteristical Lives*, 413-29. Neither Stephen Brown, in his introduction to this volume, nor Kerr, *Memoirs*, 2: 216, 416, 422-3, 430, do more than cursorily mention this work. The Newtonian Society was a student club founded in 1760, which met weekly to hear discourses on scientific subjects and debate various issues. See Wood, "Introduction," 1: viii.

> I say, I affirm, nay, if it please your reverence, I shall swear, that every order of brutes, from the grasshopper that chirps in the meadow to the lion that roars in the forest, delight as much in society, especially on particular occasions, as those of the human race. ... brutes associate together from far more disinterested motives than men. Why do men flock to large cities as bees to a hive? You, and all the orthodox, will answer, Because mankind are social animals. But I say, because they are rapacious animals. They come together in order to trick, cheat, and prey upon each other. Of all animals men are unquestionably the most unsocial. ... vanity, debauchery, and selfishness, are the true sources of ... every kind of human society. Not so the beasts of the field ... a single horse in a meadow, or a single sheep on a mountain, is not half so gay, frolicsome, or happy, as if some hundreds of their kind fed together, although by that means they may even be forced to live on short allowance. Hence horses and sheep are more social animals than the so much boasted lords of the creation.[7]

This was an implicit subversion of the law of unintended consequences, made famous in the eighteenth century through the social theories of Bernard Mandeville and later Adam Smith. Denying any "public benefits" accruing from "private vices" was a common criticism of Mandeville's ideas. Smellie, however, here utilized this criticism in an uncommon manner, connecting it to the claim that animals were in effect more social and more moral creatures than human beings. Shortly afterwards, Smellie gave his favorable view of animals much more sustained attention in the now almost-forgotten essay *Whether Are All Animate and Inanimate Bodies Made for the Immediate Use and Conveniency of Mankind; or, is That Only a Secondary End of Their Existence?* He began this essay by claiming that it was no wonder that the doctrine outlined in its title was very popular, as it "gratifies that arrogance and presumption which so universally characterize the human species."[8] Human beings obtained many advantages from the use of animals, vegetables, and fossils, "But that the advantage derived to mankind from these creatures is the primary end of their existence, or was the motive which induced the bountiful Author of Nature to call them into being, is an opinion, not only doubtful, but contradictory to reason itself."[9] Therefore Smellie wrote of "the absurdities which flow from the supposition, That the principal intention of every created substance was solely the interest

7 From a letter to Henry Hunter, in Kerr, *Memoirs*, 1: 81-2.
8 Smellie, "Whether Are All Animate and Inanimate Bodies Made," 413.
9 Ibid., 414.

of the human race."[10] To illustrate these anthropocentric absurdities he then gave many examples of animals not meant for human benefit, such as those which lived isolated from human beings, noxious and rapacious animals, animals only lately discovered and still not examined or appropriated for human use, not to mention others probably still undiscovered, as well as insects, many of which were not useful and were even detrimental to human health; similar things might in fact also be said of vegetables and fossils, and of all the natural creation.[11]

Smellie alluded to the theory of the Great Chain of Being, which substantiated the view that everything led to human beings, whose good was the ultimate end of all other creatures. While not denying the basic veracity of this theory, he nonetheless claimed that "I shall always think myself happy in the contemplation of that beautiful concatenation of being: But I cannot admit that beings are thus linked together for the sole benefit of their chief, without any regard to their own welfare and happiness."[12] This then led to criticism of the traditional biblical outlook, derived mainly from the book of *Genesis*, that all nature was meant for human use.[13]

> It is almost needless to observe here the difference between having the superiority or dominion over any Creature, and the chief end of that Creature's being. On a contrary supposition, the life of all the lower ranks of men would be chiefly intended to gratify the avarice of their Masters. The people would be created for the sake of the King, and the whole world, for the sake of a few ambitious individuals; than which nothing can be more ridiculously absurd.[14]

Smellie's later Whig political views were already evident here. The chain of natural beings did not end with humanity but was mirrored, and in fact continued, in the human social realm. Yet in both the natural and the social-political domains this chain, though a reality, was fraught with injustice. In the former, this injustice was mainly in the unwarranted license with which human beings utilized their obvious superiority to harm animals more than human need necessitated; in the latter, the political implications were obviously Whig in the common eighteenth-century sense. For Smellie,

10 Ibid., 419.
11 Ibid., 415-22.
12 Ibid., 422-3.
13 Ibid., 423-6.
14 Ibid., 424.

throughout his life, the study of natural history was part of a general world view encompassing science, philosophy, religion, and a general enlightened democratic attitude toward the propagation of knowledge. The extension of the Great Chain of Being, and its critique, to the realm of the human political world was therefore more than just metaphorical.

However, to return to the biblical sanction of anthropocentrism, Smellie noted that "The evident meaning of every tree, fruit, and herb's being given to man for food, is no more than giving him authority to select, from the vast variety of vegetables, whatever herbs, fruits, or seeds, he finds, by experience, to be most proper for the health and nourishment of his body." God, with divine beneficence, was anxious that every element of nature should be as crowded with life as possible.[15] Human beings were undoubtedly the lords of creation, but this did not sanction an unmitigated anthropocentrism, which in essence created an unjustified and corrupting influence. Smellie therefore attempted to reconcile his anti-anthropocentrism with religion, in effect claiming that a cruel unmitigated anthropocentrism was an abrogation of religion, a deviation from the true, more ethical, and original biblical cosmology.

> Of all the inhabitants of this world, man is undoubtedly the most noble. Of course, it is reasonable that he should inherit greater privileges and advantages than creatures of inferior natures; and, likewise, that he should endeavour to render as many things subservient to his peculiar advantage, as he possibly can. The superiority of his genius, and, I am afraid, frequently the cruelty of his disposition, is the cause of his extensive dominion over the brute creation. This unlimited tyranny may perhaps have given rise to the general, though mistaken notion, That these creatures were made chiefly with a view to his interest.[16]

Similarly, the Great Chain of Being theory was also essentially correct and conformed with religious cosmology. Yet it too had been similarly misused in an excessively anthropocentric manner.

> That existence, in whatever form or degree, is a blessing to the possessor, is an undeniable truth. ... It is a maxim in Nature, That no blank should be left in the vast scale of being. A strict adherence, I imagine, to this catholic maxim, is the reason why such prodigious troops of animals, not only useless, but even noxious to mankind, are called into life. Accordingly

15 Ibid., 426-7.
16 Ibid., 427.

> we see, that, lest any void should be left, mankind are suffered to be infested, and, in many instances, actually injured, by creatures which all his ingenuity is unable to turn to any advantage. This, again, throws light upon another maxim, That all partial evil is universal good; for even man, the chief of animals, must submit to many evils and inconveniencies, rather than that the blessings of life should be denied to insects and wild beasts.[17]

By including this Leibnizian allusion to "the best of all possible worlds" and to the Great Chain of Being as essentially a component of a pre-established harmony, Smellie was ostensibly revealing what he had never in fact denied: the religious foundation of his opinions and the fact that he regarded an unmitigated anthropocentrism as a travesty of religion. Unsurprisingly, therefore, he ended the essay by asserting that the proper, anti-anthropocentric, view of the natural world could "reconcile us to many of the mysteries of Providence, 'and justify the ways of God to man'."[18]

Later, in *The Philosophy of Natural History*, his best-known and mature work, Smellie sought to propagate natural knowledge more widely in society, while connecting it to an understanding of humanity's place in the divine ordering of nature.[19] This was a continuation of the early modern attempt to reconcile the new science with the old theology, a "natural theology" which increasingly gave way before an outright materialism in scientific thought. Smellie's position was therefore a curious blend of the old and the new, of religiosity and science. Criticism of anthropocentrism was an interesting manifestation of this intellectual program. For example, he wrote:

> The magnitude of this earth, when considered with relation to man, and other animals, appears to be exceedingly great. ... But, when compared to the other heavenly bodies, the number and magnitude of which exceed all the powers of human conception, the grandeur of our earth diminishes. Instead of exciting wonder, it almost vanishes from our sight. Instead of an immense globe, it dwindles into a point, seems to occupy no space, and loses itself in the boundless regions of the universe. Considerations of this

17 Ibid., 428. The young Smellie seems to have struggled with the problem of theodicy. He discussed this in relation to some of Hume's observations, which he criticized, while at the same time in effect not denying the existence of certain evils in the world. Robert Kerr felt uneasy enough on this point to excuse Smellie's views in this case because of his then young age. See the letter to George Campbell from 1765, in Kerr, *Memoirs*, 1: 214-18, and 219 for Kerr's comment.

18 Smellie, "Whether Are All Animate and Inanimate Bodies Made," 429.

19 Wood, "Introduction," 1: xvii-xviii.

> kind are apt to depress the dignity of man, and to lessen his importance in the great scale of being; but they expand his mental faculties, and exalt his ideas concerning that inconceivable Power which first produced, and still supports, those astonishing orbs.[20]

Since the sixteenth century, the Copernican Revolution and subsequent astronomical discoveries had done much to shake the traditional belief in humanity as the divinely ordained apex of nature. Nevertheless, these did not really overturn this deeply ingrained cosmology, particularly when the practitioners of the Scientific Revolution initiated the drive for mastering nature, while often emphasizing this as a divinely sanctioned project.[21] When Smellie opposed this traditional, religiously-based anthropocentrism, he did not consider this as taking an anti-religious stance. On the contrary, he was a religious man who criticized David Hume's opinions on miracles and saw his own endeavors in natural history as enhancing the recognition of divine omnipotence as revealed in nature.[22] This attempt to reconcile novel reformative ideas with traditional religion was a common element of the mainstream of the Scottish Moderate Enlightenment. Smellie was probably influenced on these points by Joseph Addison's reflections in number 565 of *The Spectator*, where Addison, contemplating the great size of the universe, "was afraid of being overlooked [by God] amidst the Immensity of Nature." However, he found solace thinking of God's mercy toward all his creatures.[23] The views of both Addison and Smellie regarding this issue demonstrated an important characteristic of eighteenth-century anti-anthropocentrism: it was not so much intended to belittle the notion of human dignity as to enhance humility and thus improve moral conduct. In this respect, such anti-anthropocentric notions were motivated by an ultimately anthropocentric ethic.

Smellie's anti-anthropocentric opinions were evident in the many references to animals throughout *The Philosophy of Natural History*. At a time when the debate regarding the mental characteristics of animals, and the possibility of moral duties toward them, was growing in importance and

20 Smellie, *Philosophy of Natural History*, 1: 135.

21 For the changing view of humanity's place in nature in the Renaissance, see Cassirer, *Individual and the Cosmos*; and Koyré, *From the Closed World*.

22 On these various points regarding Smellie's religiosity, see Kerr, *Memoirs*, 1: 196-203, 213-18, 387-92; 2: 459-60.

23 Addison, *Spectator*, 4: 279-83 (280-1 for the quotation). Smellie studied *The Spectator* closely, particularly when he tried to emulate it in the short-lived periodical *Man of the Moon*. See Kerr, *Memoirs*, 1: 120-2, 168-9.

popularity, this in itself was not unique. However, Smellie's overt criticism of the human devaluation of animals definitely belonged to the more extreme eighteenth-century opinions on this issue. For example, he wrote how "Of all rapacious animals, *Man* is the most universal destroyer. The destructions of carnivorous quadrupeds, birds, and insects, is, in general, limited to particular kinds. But the rapacity of man has hardly any limitation. His empire over the other animals which inhabit this globe is almost universal. He accordingly employs his power, and subdues or devours every species."[24] Human beings made use of various domesticated animals. "The ox-kind, in particular, after receiving the emoluments of their labour and fertility, he rewards with death, and then feeds upon their carcasses. Many other species, though not commonly used as food, are daily massacred in millions for the purposes of commerce, luxury, and caprice. Myriads of quadrupeds are annually destroyed for the sake of their furs, their hides, their tusks, their odoriferous secretions, &c."[25] This human power extended to all types of animal species, including birds, fish, and reptiles. "Neither air nor water" could

> defend against the ingenuity, the art, and the destructive industry of the human species ... Thus man holds, and too often exercises, a tyrannical dominion over almost the whole brute creation, not because he is the strongest of all animals, but because his intellect, though of a similar nature, is vastly superior to that of the most sagacious of the less favoured tribes. He reigns over the other animals, because, like them, he is not only endowed with sentiment, but because the powers of his mind are more extensive.[26]

Animals, like human beings, were endowed with feeling ("sentiment") and mental powers, but these were no match for human intellect. Humanity's empire over the animals was not absolute, however, and some animals escaped it and even attacked human beings. Human power was not over species but only over individual animals. Man was mortal, like all creatures. "He has no influence on the universe, on the motions and affections of the heavenly bodies, or on the revolutions of the globe which he inhabits."[27] Nevertheless, Smellie approved of the basic human empire over nature as

24 Smellie, *Philosophy of Natural History*, 1: 375.
25 Ibid., 1: 375-6.
26 Ibid., 1: 376-7.
27 Ibid., 1: 377.

long as it was not morally abused. In a passage redolent with the influence of Buffon, he commented on how human beings, by uniting together in society, were transformed from helpless individuals into a species with an almost complete empire over other species.[28] This conjectural-history perspective on human-animal relations was, as we will see later on, much more innovative than old forms of discussion of animals. In Smellie it did not receive sustained treatment, yet as a translator of Buffon he was bound to be influenced by it. While Smellie objected to the cruel abuse of domestic animals, he thought that moderate consumption of meat was natural to human beings.[29] Providence approved of the preying of animals on each other, hence also of flesh eating.[30] The capture and killing of animals for scientific purposes was also justified, if done in the least cruel manner.[31] Smellie was more supportive than Buffon of the theory of the Great Chain of Being.[32] In a chapter titled "Of the Progressive Scale or Chain of Beings in the Universe," which came at the end of the first volume of *The Philosophy of Natural History* (the only one published in his lifetime; the second volume was published posthumously), Smellie placed humanity at the head of the chain, yet he did not view this as an argument for belittling nature. All creatures, even seemingly harmful, were perfect in their way and played a part in the chain. There were differences between individuals of the same animal species, as between individual humans or between nations of men. Revealing his religious outlook on the very foundation for examining nature scientifically, Smellie claimed that human beings should be content with their place in the chain and should admire the divine creation.[33]

Human beings were intellectually superior to animals. The latter had a capacity for pain, pleasure, and language, but neither articulate language, nor a conception of the future or death, nor imagination, nor the ability to develop their instinctual abilities and to convey ideas from generation to

28 Ibid., 1: 377-8.

29 Ibid., 1: 60-1, 217-19; 2: 457.

30 Kerr, *Memoirs*, 1: 158.

31 Ibid., 2: 68, 70.

32 Garrett, "Introduction." Rather than outline a specific version of this theory, Buffon opted for a more limited progression by imperceptible degrees from plants to animals, and from lesser to more developed animals. See "Analogies between Animals and Vegetables," Chapter 1 of "General History of Animals," in Buffon, *Natural History*, 2: 1-15. However, he perceived a distinct distance between the more spiritual human beings and the rest of the animals. See the remarks in "A Dissertation on the Nature of Animals," in *Natural History*, 3: 224. References to Buffon's work in this chapter will be to Smellie's translation rather than to the original.

33 Smellie, *Philosophy of Natural History*, 1: 520-6.

generation.[34] In a chapter of *The Philosophy of Natural History* titled "Of the Instinct of Animals," one of Smellie's most famous productions, he claimed that human characteristics and conduct such as avarice, superstition, devotion, and morality were all modified, compounded, or extended instincts. All faculties of the human mind also existed in animals, which were rational beings to a certain extent.[35] The wide range of human instincts constituted their superiority, and

> it seems to be apparent, that instincts are original qualities of mind; that every animal is possessed of some of these qualities; that the intelligence and resources of animals are proportioned to the number of instincts with which their minds are endowed; that all animals are, in some measure, rational beings; and that the dignity and superiority of the human intellect are necessary results, not of the confirmation of our bodies, but of the great variety of instincts which Nature has been pleased to confer on the species.[36]

Smellie's friend, the famous geologist James Hutton, developed a slightly different comparison between human beings and animals, claiming that animal instinctual behavior, though actuated by reason, was fixed and unalterable throughout life, whereas human wisdom was more capable of abstract reasoning and thus of acting toward preconceived ends and developing morality above and beyond the temporally limited concerns of animals.[37] Given this stricter division between human and animal mental properties, it is no wonder that Hutton was also more conservative in his opinion that nature was created for the sake of humanity.[38]

34 Ibid., 1: 61-3, 108-10, 144-5, 147, 149; 2: 347, 414, 422-4, 434-7, 445, 457, 459-60, 464-9, 481-2.

35 Ibid., 1: 144-59.

36 Ibid., 1: 159. This chapter was based on a discourse Smellie gave before the Royal Society of Edinburgh in 1785, a summary of which was published in the society's transactions in 1788. See *Transactions of the Royal Society of Edinburgh*, 39-45; Kerr, *Memoirs*, 2: 224; Wood, "Introduction," 1: xiv.

37 "The Morals of Mankind Compared with the Instinctive Manners of the Brute," in Hutton, *Investigation*, 3: 514-22.

38 Hutton, *Theory of the Earth*, 1: 17-18. For Smellie's friendship with Hutton, see Brown, "Introduction," viii. For similar anthropocentric notions expressed by Carl Linnaeus, see Rausing, "Underwriting the Oeconomy," 188-91. Linnaeus had attracted the attention, and criticism, of both Smellie and Buffon, on several points, although Smellie appreciated, and was influenced by, certain aspects of his scientific theories. For details, see Wood, "Introduction," 1: vii, xix-xxi. On Linnaeus and his influence, see also Worster, *Nature's Economy*, 25-55.

Smellie was much more of a theriophile. While he did not deny the basic religious anthropocentric view of human superiority, he was critical of its abuse, as an excuse for unnecessary insensitivity to the animal creation. An interest in animals, as we will see in the following chapter, was consistently expressed by Scottish Enlightenment philosophers. In his later years Smellie would undoubtedly have shared his sympathy toward animals with Robert Burns, after they became close friends in 1787 when the poet, whose affection toward animals was well-known, most famously in the poem *To a Mouse*, came to Edinburgh, and Smellie printed his poems. Both men were members of the rowdy club known as the "Crochallan Fencibles," Smellie's notorious membership in which did not help his failing social and professional prospects in his later years.[39]

A prime influence on Smellie's intellectual views was the thought of the Comte de Buffon, many sections of whose work he laboriously translated, learning French in the process. Yet Smellie was not a blind adherent to all of the Frenchman's opinions. The Scottish Enlightenment was in general relatively more conservative than the French, and it objected to Buffon's materialism and irreligion. Smellie shared these points of disapproval and also criticized Buffon's rejection of final causes.[40] While in *The Philosophy of Natural History* Smellie did not get entangled in a dispute with Buffon's irreligion, it is clear that this was a major point of difference between the views of both naturalists.[41] However, there were other issues which separated their respective cosmologies, not least of which was Buffon's much more committed anthropocentrism and his eulogy of human mastery of the forces of nature. One of the main texts in which this was evident, and which Smellie translated, was "Of Nature. First View." There Buffon wrote of the beauty of nature cultivated by human beings, who lost their right of conquest over nature when they engaged in ruinous wars which were followed by famine and depopulation.[42]

> Nature is the external throne of the divine magnificence. Man, who contemplates her, rises gradually to the internal throne of the Almighty. Formed to adore his Creator he has dominion over every creature. The vassal of heaven, the lord of the earth, he peoples, ennobles, and enriches

39 Kerr, *Memoirs*, 2: 259, 349-57, 455-7. For Burns's attitude toward animals, see Perkins, *Romanticism and Animal Rights*, x, xiii, 4, 5, 7-13, 31, 40, 49-52, 57, 60, 65-6, 75, 107, 132, 180 (for cursory references to Smellie, see 26, 117, 178).

40 Garrett, "Introduction"; Wood, "Buffon's Reception," 178.

41 Wood, "Introduction," 1: xix-xx.

42 Buffon, *Natural History*, 6: 249-63.

> this lower world. Among living beings he establishes order, subordination, and harmony. To Nature herself he even gives embellishment, cultivation, extension, and polish.[43]

The religious overtones of this passage, however, seemed no more than lip-service to his readers. Buffon quickly forsook them when, after depicting with various examples the ways in which human cultivation improved nature, he claimed that "a thousand other monuments of power and of glory sufficiently demonstrate that man is the lord of the earth; that he has entirely changed and renewed its surface; and that, from the remotest periods of time he alone has divided the empire of the world between him and Nature."[44]

The other major text where Buffon's praise of the human mastery of nature was evident, in an even more detailed manner, was the essay "Concerning that Period when the Powers of Man aided those of Nature."[45] In this essay Buffon outlined an example of a conjectural history of human cultural development which undoubtedly appealed to Scottish Enlightenment taste.[46] He enthusiastically detailed the development of human societies, who by uniting into communities first defended against the forces of nature, and then gradually gained control over it by domestication of animals, agricultural development, and astronomy. Buffon here went into great detail on how, from small beginnings, human science eventually advanced to the

43 Ibid., 6: 257-8. For the original, see "De la Nature. Première Vue," in Buffon, *Histoire Naturelle*, 12: i-xvi, at xi: "La Nature est le trône extérieur de la magnificence Divine; l'homme qui la contemple, qui l'étudie, s'élève par degrés au trône intérieur de la toute-puissance; fait pour adorer le Créateur, il commande à toutes les créatures; vassal du Ciel, roi de la Terre, il l'ennoblit, la peuple & l'enrichit; il établit entre les êtres vivans l'ordre, la subordination, l'harmonie; il embellit la Nature même, il la cultive, l'étend & la polit." Although Smellie is known to have occasionally slightly extemporized on Buffon's text, I have refrained from amending his translations, since there are no serious disparities in the passages quoted here.

44 Buffon, *Natural History*, 6: 261. For the original, see Buffon, "De la Nature. Première Vue," xiv: "mille autres monumens de puissance & de gloire démontrent assez que l'homme, maître du domaine de la terre, en a changé, renouvelé la surface entière, & que de tout temps il partage l'empire avec la Nature."

45 This essay was included in the supplementary volumes of the *Histoire Naturelle*, which Buffon sent Smellie as a present. Smellie may have translated all of the famous *Des Époques de la Nature*, in which this essay was included, although this complete translation was never published. See Kerr, *Memoirs*, 2: 121, 127, 133-4, 139-44. Smellie's translation of the original title of this text (cited below) was imprecise.

46 We will discuss conjectural history in some of the following chapters, where it will be a major topic.

modern ability to create new species of plants and animals.[47] Human beings were even capable of influencing the climate, and combating its constant tendency to cool by engaging in deforestation, the control of rivers, the use of fire, the population of geographical regions, and other forms of development of natural resources. According to Buffon the temperature of a given area was defined by the ratio between its number of people and domesticated animals, and that of its wild uncultivated vegetation, since plants created humidity and cold.[48] In order to attain such cultural achievements, peace was a necessity, yet human beings still neglected morality and science in favor of the arts of war and amusement. In a peaceful state nothing could hinder their control of nature.[49] After depicting various detailed examples of the human cultivation of plants and animals, Buffon continued and wrote how

> All these recent examples show, that it was long before man knew the greatness of his power, and that he is not yet fully acquainted with its extent: It depends entirely on the exercise of his intellect. Thus the more he shall observe and cultivate Nature, the more expedients he will discover for making her submit, and for drawing from her bosom fresh sources of riches, without diminishing the inexhaustible treasures of her fertility.[50]

All this was redolent of an anthropocentric ethic and of an outlook which was predominantly rational, promoting the scientific endeavor without according religion any significant role.[51] This conjectural-history outlook regarded the scientific utilization of nature as a central aspect of the progress of human civilization. Though Buffon's scientific activity was different from the anatomical investigations of medical animal vivisection experimenters, he regarded the whole project of what later came to be called the Scientific Revolution as a manifestation of the human process of cultural progress,

47 Buffon, *Natural History*, 9: 381-410.

48 Ibid., 9: 397-403.

49 Ibid., 9: 394-6.

50 Ibid., 9: 409. For the original, see "Septième et dernière époque. Lorsque la puissance de l'Homme a secondé celle de la Nature," in Buffon, *Histoire Naturelle*, Supplément, 5: 225-54, at 253: "Tous ces exemples modernes & récens, prouvent que l'homme n'a connu que tard l'étendue de sa puissance, & que même il ne la connoît pas encore assez; elle dépend en entier de l'exercice de son intelligence; ainsi plus il observera, plus il cultivera la Nature, plus il aura de moyens pour se la soumettre & de facilités pour tirer de son sein des richesses nouvelles, sans diminuer les trésors de son inépuisable fécondité."

51 For discussions of Buffon's views on these various issues, see Glacken, *Traces on the Rhodian Shore*, 565-81, 587-91, 655, 663-81, 704-5; Roger, *Buffon*, 228-67, 418-21; and Spary, *Utopia's Garden*, 99-154.

progress which ultimately led to the higher forms of social and political innovation. Animals for Buffon were therefore part of the general philosophy of the Enlightenment, with its irreligious bent. Much more conservative than Buffon, Smellie still saw the preoccupation with natural history as a way of comprehending the divine presence in the material world. Yet he did so while discarding the anthropocentric overtones of religious cosmology, and in this he was more original than both proponents of this traditional outlook on the one hand and advocates of the new rational science like Buffon on the other. Traditional religious anthropocentrism was seldom seriously challenged in the seventeenth and eighteenth centuries. On the contrary, this anthropocentrism perfectly served the emerging scientific endeavor to master nature. One should also remember that most eighteenth-century intellectuals received an essentially religious upbringing (one thinks for example of Voltaire's Jesuitical education). Complete atheists, contrary to a rather common misconception, were a rarity in the eighteenth century.[52] Given the growing exaltation of the scientific mastery of nature, there rarely arose a serious need for contesting the common traditional anthropocentrism. Most Enlightenment philosophers were probably unaware that they were perpetuating one of the most fundamental traditional religious doctrines – the anthropocentric cosmology which sanctified the human mastery of nature. The transformation of this cosmology from a religious to a rational-scientific philosophy eventually led to the even more potent modern ethics of utilization of nature. This was far from intentional on the part of Enlightenment intellectuals. In the pre-Malthusian age the mere notion that an environmental crisis ensuing from all this was at all possible, that nature's bounty was not infinite, would have surprised the vast majority of them. Yet we should be careful not to fall into the historically anachronistic temptation to judge previous eras by present circumstances. This is wisdom in hindsight, and distorts a proper understanding of the eighteenth century in its own terms.

It is against this background of historically persistent anthropocentrism that Smellie's originality becomes evident. More moderate anti-anthropocentric philosophical argumentation, and theriophilic concern for animals, were not uncommon in the seventeenth and eighteenth centuries. Yet by and large such views criticized human pride and presumption specifically in order to humble human beings and ameliorate their conduct. This became clear when the interests of animals collided with those of humans, in which case even the most theriophilic philosophers forsook their concern for animals.

52 Kors, *D'Holbach's Coterie.*

This, as we have already seen, was evident in the thought of Montaigne, Rousseau, Bentham, and even in the fictional views expressed in *The Turkish Spy*. Smellie also subscribed to similar anthropocentric limits to his theriophily. However, in contrast with these other prominent theriophiles, his argumentation for an anti-anthropocentric attitude toward animals, and nature in general, was more sustained and philosophically systematic. His originality was highlighted particularly in relation to the predominantly anthropocentric outlook of eighteenth-century science, with which he was so intimately familiar, much more so than most other theriophiles. Stephen Brown, a prominent authority on Smellie, has claimed that he "began his intellectual life in natural history as a conservative thinker and emerged only gradually as a controversialist."[53] This may have been true regarding his social standing within the conservative Edinburgh scientific establishment, his political opinions, and the main body of his scientific work, but it was definitely not true regarding his anti-anthropocentric ethics, which he developed already in his youth. It was also a measure of Smellie's intellectual independence, that even concerning one of the most important influences on his scientific studies, the works of Buffon, he remained discriminatory even while appreciative. This independence also mirrored the fact that his scientific endeavors were ultimately much less significant than Buffon's. Yet this fact is less important from our perspective here. In expressing his anti-anthropocentric views he was in a sense prefiguring the poetical sensitivity to nature of his friend Burns. In many ways his was a more straightforward anti-anthropocentrism. Yet he lacked the genius of his poet friend, and it was not his lot to have a lasting philosophical influence. In between Buffon and Burns, he was consigned to a position of mediocrity. However, as we have already noted in relation to *The Turkish Spy*, the history of ideas is often illuminated just as much by less-famous figures as by canonical ones. The very fact that certain intellectual ideas are able to surface at certain historical points is historically pertinent.

Smellie's lack of influence in his own time was a measure both of his low social standing, but also, more significantly, of the non-conformist nature of his criticism of anthropocentrism. However, the fact that he could outline such an original anti-anthropocentric philosophy in late eighteenth-century Edinburgh, one of the most important centers of Enlightenment philosophical and scientific thought, has real historical significance. It demonstrates that the tension between the anthropocentric ethics of mastering nature, and its potential for moral abuse, was already perceptible at this early stage

53 Brown, "William Smellie and Natural History," 193.

of the rise of modern industrial civilization. We at times tend to view our own contemporary concern with environmental ethics (itself a modern term) as an original intellectual contribution of our times which has arisen in response to the current environmental crisis. However, investigating Smellie's philosophy helps realize that this is far from being the case. The intellectual concern with anti-anthropocentric ethics developed in tandem with the emerging modern anthropocentric scientific culture. While scholars have given significant notice to the latter, the former has received much less attention. Yet anti-anthropocentric arguments have long been a part of western thought and in the eighteenth century they assumed a distinctly modern form as they became directly attuned to the rising rational scientific outlook. William Smellie was one of the most original and forceful proponents of this new type of anti-anthropocentrism. He was not the greatest of eighteenth-century intellectuals, but in this context he was an important figure. Like his contemporaries, he believed in the central social utility of promoting science, and would probably have been surprised at the severity of the modern environmental crisis. Yet from our later modern perspective, with its recognition of environmental problems, his sensitivity to the immorality of an unmitigated anthropocentrism assumes a significance which it did not have at the time he outlined his views. It demonstrates that Enlightenment thought already recognized the ethical complications which its advocacy for utilizing nature and animals entailed.

6 John Gregory and Scottish Enlightenment Views of Animals

William Smellie's views were relatively extreme, yet an interest in animals was expressed by most of the figures of the Scottish Enlightenment. The constitutive contribution of Adam Smith will be the subject of one of the following chapters. Here we will outline a more synoptic view, although we will give particular attention to the physician John Gregory, best known for his contributions to the early development of medical ethics.[1] Gregory, like most other Scottish contemporaries, was more conservative in his attitude toward animals, but at the same time was not completely unreceptive to the idea of moral obligations toward them, even if less so than Smellie. The views of eighteenth-century Scottish philosophers were influenced by, and reflected, general contemporary European intellectual developments, although the Scots were often at the forefront of philosophical innovation. They did exhibit certain typical traits, however, notably those influenced by the thought of Rousseau, particularly the latter's primitivistic notions regarding the stages of development of human society. The Scots, however, most often subverted Rousseau's ultimate aim; instead of utilizing conjectural-history analysis of social development to criticize advanced civilization, they used it to commend it.[2] We will take a closer look at conjectural history in the following chapters. At this stage it is enough to note that it was concerned with theoretical suppositions about human social and cultural development in the early pre-literate stages of history. The comparison of human beings with animals was an integral part of this theorizing. For our purposes, its main objective, the clarification of human nature, is less important. While animals were usually utilized mainly as similes for elucidating human issues, occasionally a consideration of them in themselves was revealed in such discussions. This was true in the case of Gregory. His version of this debate in his book *A Comparative View of the State and Faculties of Man with Those of the Animal World* followed other Scottish philosophers in according animals only a limited amount of attention, especially when discussing the primitive, savage, sensual, and instinctive elements of human nature. Yet

1 For his life and work see McCullough, *John Gregory*; Haakonssen, *Medicine and Morals*; and *John Gregory's Writings*. For a further interesting work he wrote which is irrelevant here, see Gregory, *Father's Legacy*.

2 For Rousseau's influence on Gregory, see Wood, "Natural History," 91-4, 106-8.

in this context he devoted more attention to the animals *per se* than many of his contemporaries.

While there is no clear indication that Gregory opposed a benign treatment of animals, he clearly regarded human interests as paramount. This was evident from his support of animal experimentation. In his *Observations on the Duties and Offices of a Physician and on the Method of Prosecuting Enquiries in Philosophy* he wrote:

> In order to illustrate the human physiology, a knowledge of the comparative anatomy of some animals, that most nearly resemble man, is extremely requisite. Several of the most important discoveries in the animal œconomy, have been made or illustrated by observations or experiments first made on brutes, many of which it was impossible to have made on the human subject, e.g. the experiments relating to the circulation of the blood, respiration, muscular motion, sensibility and irritability of different parts of the body, and the effects of various medicines. The instincts of brutes have often given the first hint of valuable remedies, and might throw light on the subject of regimen, and the cure of diseases, if they were properly attended to. At the same time it must be acknowledged, that the comparative anatomy of other animals has often led into great mistakes, by too hastily applying it to the human body.[3]

Whether Gregory actually performed vivisections himself is unclear from this passage, although given his medical training this was highly likely. Even his cautionary remark about the dangers of vivisection was based on human, not animal, interests. His awareness of the problematical aspects of comparing animal with human physiologies was, as we have already seen, prefigured by scientists in the previous century.

Gregory did, however, exhibit at least some adherence to certain moderate theriophilic principles. The concurrence of these with support of animal experimentation should come as no surprise, since we have seen it appear in more committed theriophiles such as Rousseau and Bentham. Perhaps a closer comparison could be made with Christiaan Huygens, for whom also animal experimentation had been more than a theoretical interest. Gregory's

3 *John Gregory's Writings*, 121. For similar quotations, see ibid., 78, 190; and McCullough, *John Gregory*, 250. For Gregory's interest in comparative anatomy see also McCullough, *John Gregory*, 45. The remarks in ibid., 248-9, and McCullough, "Hume's Influence," 385-6, imply that Gregory was perhaps not insensitive to animal suffering. But even if true, this sensitivity was very limited compared with his support of animal experimentation.

main ideas regarding this topic were expressed in *A Comparative View*, which was based on discourses he delivered to the Aberdeen Philosophical Society, also known as the "Wise Club." This work was quite popular at the end of the eighteenth century in Britain and other countries.[4] Gregory outlined the general topics of the book in the preface. He observed, in rather common primitivistic and theriophilic tones, several stages of development of human culture. Humanity in its savage state was in certain aspects in worse condition than other animals. Human beings had superior faculties, but not the same level of instinct, and this led to unhappiness. While people in this savage stage of development had more physical qualities than those in more advanced states of society, they lacked the same levels of the nobler and more distinguishing principles of human nature. In the savage state they exhibited courage, but not the quality of forgiveness which made courage a virtue. Most of their time was devoted to the quest for food or to supine sloth.[5]

At a later stage of human social progress, people possessed the physical qualities and all the animal functions in their full vigor. Such things as religion and patriotism appeared at this stage, "in which Nature shoots wild and free." Yet at the same time it checked the influence of rationality, which led to the combination of imagination controlled by reason. The main problem with this state of society was one of communication, of transferring knowledge between individuals.[6] In any event it did not last long. The rise of communication also led to commerce with other nations; new wants appeared, but also new knowledge. These developments therefore had both good and bad consequences. Very extended commerce led to moral corruption if not counter-balanced in some fashion, for example by similar developments in neighboring countries.[7] At a more advanced stage of human development, the wants created by luxury gave rise to powers of invention. These in turn encouraged elegant arts and natural principles of taste, all of which led to innocent pleasure. The investigation of nature

4 On the Aberdeen Philosophical Society, see the editorial comments in Ulman, ed., *Minutes*, 11-63; and for Gregory's discourses concerned with *A Comparative View*, see ibid., 12, 27-8, 61, 87, 236. Other members of the society who proposed similar questions related to animals for deliberation at some of their meetings, were James Dunbar and James Beattie (on both of whom see below), as well as David Skene, who once proposed deliberating whether animals had souls, and if so in what these differed from human souls (see ibid., 194). The minutes regrettably do not contain detailed accounts of these discussions.

5 Gregory, *Comparative View*, 1: iv-vi.

6 Ibid., 1: vi-viii.

7 Ibid., 1: xiv-xix.

and the development of manners were also advanced at this stage, but all of these improvements ran the risk of occasionally being perverted and applied inappropriately.[8]

Gregory proposed that the various advantages of these different stages of human development might be united, thus making life more comfortable and happy. This might be a difficult goal at the broad social level, but it was much more attainable for individuals. Thus a person might enjoy the benefits of advanced society without suffering from its possible defects. And it was the elaboration of this principle which was Gregory's professed aim in his book.[9] In this respect, he was characteristic of the moderate attitude of much of the Scottish Enlightenment. He aimed for the middle, moderate ground, between nature and progress. This was a far cry from the more blatant primitivism of Montaigne or Rousseau. While Gregory was well-acquainted and in some respects influenced by Rousseau's philosophy, particularly regarding education, in this context he differed from him. Consequently, he was also much less of a theriophile than Rousseau, whose own theriophily, as we have seen, was not unlimited. Gregory's main objective was to advance the betterment of the human condition, but to do so he required a comparison of human and animal qualities. A Cartesian-style mechanistic view of life, which in any event was already out of fashion in the second half of the eighteenth century, was therefore out of the question, and the connection between the mind and the body, even if it was difficult to elucidate its precise nature, was essential for explaining human action.[10]

Understanding human nature required comparing human beings with animals, and Gregory regarded as "very weak and foolish" the opinion that this denigrated human dignity.[11] It was at this point that he began to develop his particular views on animals, while attempting to walk a fine line between pointing to similarities which human beings shared with animals and retaining the clear superiority of the former. Most early modern theriophiles, even the more extreme among them, rarely denied human mental superiority to animals, if only in degree, and Gregory was no exception. He consistently regarded the study of humanity as the most useful and entertaining branch of natural knowledge.[12] This was

8 Ibid., 1: xix-xxi.

9 Ibid., 1: xxi-xxii.

10 Ibid., 1: 7-9; *John Gregory's Writings*, 60, 120-1, 128, 136, 142, 188-90, 195, 200-1, 216.

11 Gregory, *Comparative View*, 1: 9-11.

12 *John Gregory's Writings*, 80, 128, 157, 200-1, 238; McCullough, *John Gregory*, 247, 258.

impossible, however, without studying animals as well. Referring to the theory of the Great Chain of Being, he noted that "no one link of the great chain can be perfectly understood, without the knowledge, at least, of the links that are nearest to it."[13] All animal species possessed distinct qualities, but also many other ones which they all shared. However, "Man is evidently at the head of the Animal Creation ... If he is not the only Animal possest of reason, he has it in a degree so greatly superior, as admits of no comparison."[14] Here Gregory departed from the Great Chain of Being, claiming that the gradation between animal species was irrelevant for humanity, since "There is an infinite distance between the faculties of a Man, and those of the most perfect Animal; between intellectual power, and mechanic force; between order and design, and blind impulse; between reflection, and appetite."[15] Here humanity was considered as above and beyond the Great Chain of Being. Gregory, however, was inconsistent on this point. He seemed to simultaneously affirm and deny humanity's place in this chain. This inconsistency was not unusual in early modern thought. According to early modern primitivistic notions, human beings did indeed have the potential of superiority over the animals, but they did not always realize this potential. In light of the general Enlightenment ideal of bettering the human condition through education and moral and social improvement, which Gregory shared, his views on the Great Chain of Being would therefore not have been considered inconsistent.

In attempting to consider animals as exemplars for human conduct, Gregory was therefore continually on guard to protect human superiority. Animals governed each other by force or cunning and did not utilize reason to advance mutual assistance. They did not exhibit true feelings of superiority or subordination, and this was not disproved by examples from the animal world such as the conduct of bees or birds, examples which were customarily utilized by more committed theriophiles than Gregory.[16] The lack of animal language was not explained physiologically, since some animals were able to pronounce words. Rather, it was due to the lack of a "regular train or order in their ideas." The "works" of animals were uniform, while each individual human being thought and acted in a unique manner. Only in the animals most connected to humanity, mainly dogs and horses, did one

13 Gregory, *Comparative View*, 1: 11.

14 Ibid., 1: 11-12.

15 Ibid., 1: 12-13.

16 Ibid., 1: 13.

find exceptions to this. Gregory claimed that all animals expressed pain and pleasure by cries and various bodily motions, a perception which tacitly conceded the theriophilic view that they were sentient. He also claimed that only human beings laughed or cried, since these were emotions known only to them, and not to animals or even to human infants at a very early age. Also unique to humanity were the enjoyments of the imagination, science, the free arts, and particularly a moral sense, and the joy arising from religion and from the various intercourses of social life.[17]

Gregory was interested in the points of human superiority not for their own sake but mainly for how far they could be cultivated so as to render human beings more happy, wise, and distinguished.[18] There were also certain things in which animals were superior to human beings. Many animals had more acute external senses, although these were usually useless from a human point of view. What was important was to determine whether there were elements of animal superiority in relation to which human beings might be able to aspire at least to equal them.[19] Wild animals, if they were not killed, usually reached the ends of their natural lives, while human beings (and domestic animals) usually did not attain this and suffered a "thousand nameless ills" resulting from superior human faculties. Gregory did not specify precisely what these ills were, but the implication was clear – it was human conduct in its most advanced forms which led to the shortening of human life. If this was the design of nature, one needed to accept it, but if it was an adventitious and unnatural state of affairs, one could try and find out how such conditions arose and how they might be remedied. Thus, on the basis of a moderate primitivism, Gregory sought to improve the human condition.[20]

Gregory proceeded to explain the principles of animal behavior. The basic one was instinct, which led to the shortest and most effectual courses of animal action.[21] "Instinct" was the traditional early modern term utilized to explain animal actions, in most instances without resorting to the possibility of animal reason. Smellie, as we have seen, accorded it particular scientific attention. Gregory did not accept the clear differentiation between animal instinct and human reason. He claimed that human beings also had their share of instinct. Moreover, "as far as it extends, it is a sure and infallible

17 Ibid., 1: 14-16.
18 Ibid., 1: 16.
19 Ibid., 1: 16-17.
20 Ibid., 1: 17-19.
21 Ibid., 1: 19-20.

guide," even though the depraved and unnatural condition of humanity often stifled its voice. Reason was a weak principle in human beings and usually an unsafe guide compared with instinct. It was intended to determine the causes of things, the consequences of actions, and the best ways to attend certain ends. It was therefore meant to be a check upon the instincts, tempers, passions, and tastes. Yet these still remained the impelling principles behind human action, and "life, without them, would not only be joyless and insipid, but quickly stagnate and be at an end."[22]

Some of the animal advantages over humanity were apparent to a large extent in those human beings just above the animals, who were guided by instinct alone. These human beings were therefore strangers to the noble attainments their nature was capable of, but also to the many miseries common among "their more enlightened brethren of Mankind."[23] This was not exactly a Rousseauistic concept of "noble savages," but Gregory did imply that there was room to learn from primitive human societies. It was important to inquire what were the natural human instincts and to separate them from negative cravings. When this was complicated, one had to check the analogous instincts of animals, and particularly "those of the savage part of our own species." This was very difficult, however, since there was no class of human beings completely governed by instinct, nature, or common sense. Both the most barbarous and the most polished and luxurious of nations were differentiated by their manners and deviated from nature. They were all equally guided by reason, and variously perverted by prejudice, custom, and superstition. Yet the hand of nature was nonetheless distinguishable in savage nations, which exhibited just and acute signs of reasoning similar to those found in enlightened nations.[24]

There were certain aspects of the comparison of animals with human beings which were destined for failure, since animals were guided solely by instinct, whereas human beings were also guided by other principles, "particularly by the feeble and fluctuating principle of Reason." It was difficult to determine the line between the natural and artificial condition of humanity and to ascertain the clear border between the spheres of instinct and reason. Nevertheless, there existed certain conditions which all human beings agreed were contrary both to nature and to reason. In each particular circumstance, human beings were therefore obliged to act according to the guidance likely to lead to the most perfect and lasting happiness. Where

22 Ibid., 1: 20-1; McCullough, *John Gregory*, 112, 200.

23 Gregory, *Comparative View*, 1: 21-2.

24 Ibid., 1: 22-3.

the voices of nature and instinct were particularly clear and explicit they were to be followed. But where these voices were silent or doubtful, then one had to attend to the natural analogy in other animals, not as an absolute rule but as a guiding principle. In so doing, the ultimate appeal was to be to cool and impartial experience.[25]

At this point Gregory proceeded to utilize the human-animal comparison in a discussion of education.[26] From here on the book concentrated on various spheres of human culture which exhibited superiority to the animal condition or, more precisely, the possibility of such superiority, subject to certain conditions. Animals as a separate subject of interest almost ceased to exist at this point, since Gregory of course had already set the stage for concentrating on the specifically human issues which interested him. The first subject he addressed was child-rearing and education, in a manner obviously much indebted to Rousseau and quite reminiscent of modern notions emphasizing closeness to nature.[27] This discussion exemplified, according to Gregory, the calamities which befell the human race and which could be avoided by following nature. In childhood, instinct was the only active principle of human nature, and therefore in this context the analogy with animals was most complete. When the more unique and superior human faculties began to develop, this analogy was less complete. However, the source for the weak and sickly habits of human beings was in infancy, and this influenced the evils of body and mind which appeared in later stages of life.[28]

The decline of human life, the gradual decay of the more humane and generous sentiments, and that of "all our boasted superior powers of imagination and understanding" led eventually to human beings reaching a state more helpless and wretched than that of any animal. This situation was undoubtedly "the most humbling consideration to the pride of man."[29] The early modern theriophilic comparison of human beings with animals was usually meant to humble humanity, with the ultimate aim of ameliorating human conduct. This did not necessarily entail moral consideration of animals *per se*. For Gregory such ethical consideration was unimportant. He was interested solely in the human advantages rising from the comparison of human beings with animals. The human deterioration he depicted was

25 Ibid., 1: 24-7.
26 Ibid., 1: 27-30.
27 Ibid., 1: 30-107.
28 Ibid., 1: 104-5.
29 Ibid., 1: 106-7.

therefore, according to his view, not inevitable and was preventable by living more naturally. There was no doubt that human art could protract life beyond the limits ordained by nature, but it was more important to consider the means of enjoying life so long as it existed.[30]

According to Gregory, "The advantages, which Mankind possess above the rest of the Animal Creation, are principally derived from Reason, from the Social Principle, from Taste, and from Religion." The contribution of these things to make human life more happy and comfortable was the main concern of the rest, the major part, of the book, which therefore contained very little direct references to animals.[31] The discussion of reason still included some such remarks. Gregory considered reason a blessing only when it was properly applied. Nature had given humanity "a variety of internal Senses and Tastes, unknown to other Animals." These were to be cultivated so that they became sources of pleasure. This necessitated culture, "the peculiar province of Reason," without which there was little happiness in life.[32] In general, Gregory's discussions of child rearing and education, reason, taste, and religion, were all actuated by a search for moderation. Human beings needed to try and seek the middle ground between the dictates of nature and instinct, which the comparison with animals helped realize, and those of the more sophisticated elements of life unique to humanity. It was the middle ground between these which Gregory constantly advocated, as part of his philosophical views aimed at the improvement of the human condition. This was a typical Scottish Enlightenment outlook. Therefore his theriophily never moved beyond very clear moderate bounds. Yet, considering other early modern attitudes toward animals, there does not seem any reason why Gregory should not have included in *A Comparative View* at least some remarks regarding the need for an ethical consideration of animals. These would not have damaged his general anthropocentric outlook, which was ultimately common to various degrees among most early modern theriophiles. The only explanation seems to be that Gregory did not consider this a particularly important issue. Of course, beyond his advocacy of animal experimentation, he did not seem to condone any wanton cruelty to animals. Yet one might have expected a more active moral stance regarding animals, particularly since even a casual glance at some of his medical ethical writings reveals a person who was highly sensitive, benign, and enlightened.

30 Ibid.

31 Ibid., 1: 109; also 1: 109-72; 2: 1-208.

32 Ibid. 1: 109-11; *John Gregory's Writings*, 97.

This omission is even more glaring since Gregory was undoubtedly familiar with the early modern British tradition of moderate sensitivity to animal sentience. This was especially evident among his Scottish contemporaries. In particular, he would have been familiar with the views of David Hume. Modern scholars disagree about the extent and nature of Hume's philosophical influence on Gregory, although not about Gregory's dissatisfaction with Hume's irreligiosity. Yet it is clear that they were on friendly personal terms and that Gregory was quite familiar with Hume's general philosophical opinions.[33] Hume's attitude toward animals has been widely studied, although scholars have failed to reach an agreement on the fine points of this issue.[34] Nevertheless, his broad attitude toward animals was not unclear. He stated emphatically that no truth appeared to him "more evident, than that beasts are endow'd with thought and reason as well as men. The arguments are in this case so obvious, that they never escape the most stupid and ignorant."[35] This did not mean that animals were spiritually or mentally equal to human beings. On the contrary, Hume made a point of the superiority of human reason, and consequently of human moral sentiments. But animals did exhibit thought and feelings similar, although not equal, to those of human beings, and he implied that he did not regard as sufficient the explanation of animal actions as guided by instinct. Human beings were in a commanding position over the animals, and although they did not owe them strict justice, they were bound by humanity to treat them gently.[36]

Hume's theriophilic views were therefore more emphatic than those of Gregory. Gregory's use of the concept of instinct, for example, was very flexible, and in many instances it was tantamount to meaning reason, although undoubtedly far inferior to human reason. His differentiation between human and animal reason was more distinct than Hume's. Perhaps this was part of the reason for his disinterest in moral obligations toward animals. Undoubtedly, his general philosophical views were less original than Hume's, and more in tune with popular sentiment. Neither man developed his theriophilic viewpoints to any real anti-anthropocentric extent similar

33 McCullough, *John Gregory*, 83-97, 141-5, 308-9, and passim; McCullough, "Hume's Influence"; McCullough, "John Gregory's Medical Ethics." Compare Haakonssen, *Medicine and Morals*, 70-4; Klemme, "[Review]."

34 Bradie, "Moral Status," 42-4, 48; Tranöy, "Hume on Morals"; Seidler, "Hume and the Animals"; Baier, "Knowing Our Place"; Clark, "Hume"; Arnold, "Hume"; Nuyen, "Hume"; Pitson, "Hume."

35 Hume, *Treatise*, 176.

36 Ibid., 176-9, 259, 324-8, 397-8, 448, 466-9, 610. Also of interest are the remarks in Hume, *Inquiry Concerning Human Understanding*, 165-8.

to Smellie, but Gregory was more committed to a clear anthropocentric philosophy. His interest in animals was systematically subjected to a concern with human affairs, and he made this clear toward the end of *A Comparative View*, where he wrote of

> The advantages arising to Mankind from those faculties, which distinguish them from the rest of the Animal world; advantages which do not seem correspondent to what might be reasonably expected from a proper exertion of these faculties, particularly among the few who have the highest intellectual abilities, and full leisure to improve them.[37]

Modern scholarship has given relatively little attention to the view of animals of other eighteenth-century Scottish philosophers besides Hume, yet these were also undoubtedly familiar to Gregory. The conspicuous attention to the ethical treatment of animals espoused by Francis Hutcheson was particularly important. Hutcheson distinguished sharply between the intellectual and spiritual characteristics of human beings and animals. The latter had senses, appetites, and lower instincts dependent on bodily organs. In contrast to human beings, their sexual desire was limited to sensual pleasure and was not connected to mutual love and united interests. They cared for their offspring for a much shorter period of time than did human beings, and their social impulse in general was more restricted. More importantly, human beings surpassed them not just intellectually but particularly by their variety of pleasure, their honor, sympathy, sense of past and future, and especially their moral sense. Animals were capable of some happiness and misery, and they could move human compassion by their suffering, yet human beings were capable of incomparably more happiness and misery.[38] Despite some moral hesitancy, Hutcheson regarded divine revelation as affording raising animals for human consumption. Animals, so he claimed, even gained from this the right to propagate and to die quickly. Perishing was a greater evil for the superior human beings, and a prudent and compassionate human dominion over the animals was beneficial to both sides. Nevertheless, people approved of relieving animal suffering, unnecessary cruelty toward them was wrong, their moral capacity was similar to human infants, and cruelty toward them might produce similar behavior toward human beings.[39] Hutcheson even went so far as to claim that

37 Gregory, *Comparative View*, 2: 206-7.

38 Hutcheson, *System of Moral Philosophy*, 1: 2-3, 29, 33-5, 309, 311-16.

39 Ibid., 1: 311, 313-14, 316-17.

> brutes may very justly be said to have a right that no useless pain or misery should be inflicted on them. Men have intimations of this right, and of their own corresponding obligation, by their sense of pity. 'Tis plainly inhuman and immoral to create to brutes any useless torment, or to deprive them of any such natural enjoyments as do not interfere with the interests of men.[40]

Aaron Garrett has accorded particular historical significance to Hutcheson's advocacy for animal rights.[41] While Hutcheson's were no doubt among the more emphatic eighteenth-century views of this topic, they were not absolutely unique, as the example of Smellie proved.

Animals were an object of at least passing interest to other Scottish philosophers, even if most of them were not particularly concerned with the moral aspects of this topic. Henry Home, Lord Kames, for example, remarked that "reason and experience have little influence on brute animals," who "have many instincts that are denied to man; because the want of them can be supplied by education."[42] Adam Ferguson went into greater detail in expressing the rather common perception of diminished intellectual and spiritual capacities of animals compared to human beings. For example, he claimed that animals were capable of enjoyment and suffering, affection for their young, sociability, expression, communication, and perhaps even intelligence. Nevertheless, human beings were more sociable, and only they were capable of such things as speech, versatility of works resulting from causal reasoning, imagination, religion, moral judgment, gratification of fancy, and the external effects of historical progress which proved intelligence.[43] Yet Ferguson also made the customary theriophilic observations, meant for human amelioration, about the singular human qualities enabling human beings not just to rise above the animals, but also to sink below them. Imperfect human reason was more prone to error than animal instinct, and human beings were accountable for their intelligence and freedom.[44] Dugald Stewart, in contrast, differentiated more sharply between animal instinct and human intelligence. Animals achieved a certain sagacity through experience, yet the moral and intellectual principles of human beings

40 Ibid., 1: 314.

41 Garrett, "Francis Hutcheson." For briefer considerations see Harwood, *Love for Animals*, 133, 161-4; Thomas, *Man and the Natural World*, 175, 179, 298; and Bradie, "Moral Status," 34-9, 49.

42 Home (Lord Kames), *Sketches*, 1: 5-8; Thomas, *Man and the Natural World*, 28, 68, 136, 208.

43 Ferguson, *Principles*, 1: 11-61.

44 Ibid. 1: 32, 54-5. See also Harwood, *Love for Animals*, 133; Thomas, *Man and the Natural World*, 292; Bradie, "Moral Status," 45, 49.

differed in kind and not in degree from those of animals. Therefore, while Stewart implied an objection to the beast-machine theory, writing toward the end of the era of the Enlightenment, he also basically rejected the early modern popularity of the Great Chain of Being.[45]

Nevertheless, toward the end of the eighteenth century there was a rise in emphatic theriophilic writing, evident in articulated defenses of benign treatment of animals such as those by Humphry Primatt and Thomas Young.[46] In Scotland this extreme type of theriophily was represented by John Oswald, in his book *The Cry of Nature; Or, An Appeal to Mercy and to Justice on Behalf of the Persecuted Animals*.[47] This was not a very systematically-argued book, but it presented an impassioned defense of the need to treat animals humanely. On an even more singular note, it included a rare attempt to include the ethical treatment of animals as part of the ideals of the French Revolution. This connection between political and animal-rights principles was still quite novel even toward the end of the eighteenth century. Oswald, however, was not a mainstream figure of the Scottish Enlightenment and was unrepresentative of the more conservative views of his contemporaries.

The mainstream Scottish Enlightenment was far from such extreme notions. The important philosopher Thomas Reid, Gregory's cousin, demonstrated a serious interest in the animal issue. He gave a detailed exposition of the common claims of differences of both kind and degree between animals and human beings. Appetite was a key element governing animal actions. Animals exhibited limited degrees of such qualities as sagacity, affection for offspring, courage, anger, shame, pride, emulation, gratitude, desire of power, and even esteem of knowledge. They had memory, but not a sense of temporal intervals, belief which was not grounded on evidence, emulation, but mainly of a physical type, as well as some degree of social intercourse, though not the ability to give testimony, to lie, to promise, or to enter into a contractual obligation. They did not exercise judgment or reason. They were devoid of morality, virtue, or vice and were not subject to moral approbation or disapprobation. They were subjects of discipline but not of law. They exhibited resentment, but not deliberate as in human beings. They did not differentiate between hurt and injury, since this was a rational act connected to a conception of justice and to a moral faculty. Unlike human beings they were unable to plan for the future or devise means to attain certain ends. They were also incapable of such things as speech,

45 Stewart, *Outlines*, 70-3.

46 Primatt, *Duty of Mercy*; Young, *Essay*.

47 Oswald, *Cry of Nature*; Wolloch, *Subjugated Animals*, 62-3, 126.

government, conscience, politics, or theology. They lacked the most noble faculties endowed by God to humanity, mainly those which made people moral and accountable beings.[48] Nevertheless, Reid also demonstrated a distinct theriophilic morality. The world, he noted, was also created for animals and their entertainment. It was because of their inferiority compared to human beings that they were considered the latter's servants. Animals which served human beings deserved wages and were not to be neglected. If God enjoyed communicating happiness to the much lowlier human beings, then the latter should emulate this benignity by behaving in a similar manner toward each other, and even toward animals.[49]

Reid, like Gregory, was a member of the Wise Club in Aberdeen, where other members of the club also expressed similar views on animals, although usually devoid of moral commentary. James Dunbar revealed the typical eighteenth-century Scottish mélange of conjectural history and moderate theriophily. He claimed that human beings were superior to animals, although not physically, and not always to the same degree, as the example of some primitive peoples proved. Animals also had inferior abilities such as a limited degree of reason, imitation, invention, recollection, foresight, and the ability to infer the future from the past. Yet they could not react emotionally and imaginatively to their environment, and their language was limited compared to human language, which was historically progressive. Nevertheless, human distinctness also entailed certain risks. Even the most rapacious animals refused to harm members of their own species, yet this happened among human beings. Clearly demonstrating the influence of Rousseau, Dunbar claimed that the improvements leading to human advancement also caused feuds and animosity; only at an intermediate stage of human development did a short Golden Age occur.[50]

On one occasion Dunbar proposed a question for deliberation at the Aberdeen Philosophical Society, regarding how to differentiate between the operations of instinct as opposed to reason and sagacity in animals.[51] Gregory's protégé James Beattie also once proposed a question for deliberation at the society, regarding the origin and extension of human authority over the animals.[52] Yet, despite showing some concern for animals in apparently opposing hunting, Beattie nonetheless expressed rather conservative

48 Reid, *Works*, 1: 359-60; 2: 544-79, 596-7, 665; Bradie, "Moral Status," 44-7.

49 Reid, *Practical Ethics*, 204-5, and also 378-9.

50 Dunbar, *Essays*, 11-19, 28-33, 43-6, 63-100, 342-4, 348-9.

51 Ulman, *Minutes*, 42, 198.

52 Ibid., 193.

views on their inferiority to human beings. He claimed they had memory, but no recollection or active remembrance. Only human beings had speech, invention, science, morality, contemplation, politics, and religion. Beattie's conservatism and religiosity probably also contributed to his apparent rejection of primitivism and the theory of the Great Chain of Being.[53] Expressing a sentiment which most of his contemporaries would have shared, he exclaimed: "Let us hence learn to set a proper value on the dignity of the human soul; and to think of its intellectual faculties as inexpressibly superiour, both in kind and in degree, to those of the animal world."[54]

From all of the above examples it is evident that Gregory's views on animals were a particularly representative and detailed example of the Scottish Enlightenment's predilection for mild theriophilic views commingled with conjectural history. This fitted well with the Enlightenment ideal of human social and moral amelioration. Gregory's version of this Enlightenment plan was clear. Human beings had qualities which potentially enabled them to rise to intellectual and social attainments impossible for the mentally more limited animals, yet these qualities could not be taken for granted. They had to be cultivated, and only a small number of people enjoyed both the characteristics and the advantageous circumstances which afforded this refinement. Such social and spiritual attainment required a moderate synthesis of cultural sophistication with common sense natural behavior. The animal world was chiefly to be discussed for its ability to afford exemplars for the latter ingredient. Animals in themselves, however, were not very important.

Gregory has become somewhat of an esoteric figure from a modern perspective. Yet in his own era, although not truly famous, he was highly respected, and some of his writings were both popular and influential. The opinions described in *A Comparative View* were far from unusual or revolutionary, but for that very reason they were representative of popular contemporary notions. Early modern theriophily was primarily a literary device utilizing animals as exemplars for human conduct. The more anti-anthropocentric, moral aspects of theriophily were the province of relatively few people. Gregory in this respect was much more representative of the Scottish Enlightenment view of animals, and indeed of the eighteenth-century view of them in general, than Smellie. It should be emphasized, however, that Gregory's type of discussion of animals, as well as the other examples noted here, were more a summation of previous early modern types of thinking

53 Beattie, *Dissertations*, 60-71, 233-9.
54 Ibid., 67.

than new types of debate about animals. The use of conjectural-history observations was in itself one of the salient features of the new type of social philosophy and the "science of man" which preoccupied Scottish intellectuals of the late Enlightenment and often surfaced in discussions of animals. But Gregory used it in relation to the old traditional philosophical debates about animals, not the more novel historiographical and economic perspectives which, as we will see in our remaining chapters, his contemporaries, not least in Scotland, were developing. Nor did he outline a unique type of theriophilic argumentation in Smellie's fashion. A further type of novelty was the extreme animal-rights philosophy advocated by John Oswald. But this was a rare phenomenon in the eighteenth century and even the nineteenth. If one looks for a new Enlightenment perspective on animals, one which still retained the traditional interest in the detailed comparison of animal with human physical and mental characteristics but at the same time developed a novel outlook on animals, then this was demonstrated most distinctly by one of the most original of the Scottish Enlightenment philosophers, Adam Smith. His contribution to this debate, as we will see later on, was specifically related to new Enlightenment historiographical and economic discourses. These eventually made the old type of discussion of animals, both philosophical and ethical, which we have been surveying so far, increasingly irrelevant. But before beginning to investigate these new types of discourse, we need to take a look at some examples of intermediate late Enlightenment discussions of animals, ones which intermingled the old and the new.

7 Buffon, Crèvecoeur, and the Limits of Enlightenment Sensitivity to Animal Suffering

By the second half of the eighteenth century, European intellectuals became much less concerned with the traditional debate regarding the capacity of animals for cognition on the one hand, or for suffering on the other. This did not mean a lack of concern with animal pain; quite the contrary. Yet particularly in the late Enlightenment, such ethical sensitivity customarily took a back seat to a growing awareness that cultural progress necessitated the utilization of animals. The debate about animal pain seemed a moot point. The animals' capacity for suffering was commonly recognized, as was the need to avoid inflicting *unnecessary* pain on them. Such pain, however, often seemed all too necessary. The discarding of the beast-machine theory notwithstanding, eighteenth-century intellectuals still faced the problem of the incongruity between the ethical consideration of animal suffering and the general overriding preponderance of human interests.

Nowhere was this outlook more evident than in late eighteenth-century historical discourses. A historical element was common to many contemporaneous discussions, not all of them belonging strictly to historical literature in the modern sense. Key to such debates was the rise, in the second half of the century, of the early type of nascent theoretical anthropology termed "conjectural history," which attempted to surmise the origins of human material and social culture.[1] We have already seen how conjectural history served the great French naturalist the Comte de Buffon in his praise of the human mastery of nature, and the constitutive place this had in the historical progress of humanity. It remains to see how this anthropocentric outlook interacted with his ethical view of animals. As with other scientists, the fact that Buffon was fascinated with the animal world does not immediately mean that his moral treatment of animals was necessarily benign. In fact, scientific praxis, even when it was not directly concerned with vivisection, often entailed cruel treatment of animals. No obvious ethical view of them

1 Meek, *Social Science*; Hont, *Jealousy of Trade*, 159-84, 364-6; Höpfl, "From Savage to Scotsman"; Berry, *Social Theory*, 61-71; Garrett, "Anthropology"; Emerson, "Conjectural History"; and for nineteenth-century influences, Palmeri, "Conjectural History."

can be inferred either from Buffon's fascination with them or from his anthropocentric cosmology.

On the one hand, Buffon indeed emphasized human superiority and the right of human beings to utilize animals. Nevertheless, he also seemed aware of the capacity of animals for suffering and of the need to consider this propensity from a moral perspective.[2] He opposed unnecessary harm to animals but regarded consumption of animal flesh as part of the natural order, helping, for example, to control the relative populations of various animal species. Human beings physically required a carnivorous diet, and Buffon opposed the primitivistic notion of vegetarian humanity in the state of nature, criticizing philosophers (mainly Rousseau) who supported this "fable, in which one treats man like an animal" (fable, où l'on emploie l'homme comme un animal).[3] Human beings destroyed more animals than all other carnivorous beasts, more due to intemperance than to real want. Yet, however great the waste created by humans, the total quantity of life never diminished, and "reproduction was born of destruction" (la reproduction naisse de la destruction).[4] Buffon regarded the human manipulation of nature as an ongoing project necessary for human progress, morally permissible both because of human innate qualities and because nature was able to sustain any amount of human manipulation. Like most of his pre-Malthusian contemporaries, he would probably have found subsequent modern notions of over-population quite surprising.[5]

The human use of animals also seemed justified because of the latter's inferiority to the former. At one point Buffon claimed, in seemingly Cartesian fashion (unusual at a time when the beast-machine theory was long discredited), that animals acted in purely mechanical fashion, while spirituality was conferred only on human beings.[6] Slightly later, however, he insisted that he did not reduce animals to insensible automatons. On the contrary, he claimed that their feelings were perhaps even more acute than human feelings, even if they did not have thought or reflection. Animals were conscious of their present existence, yet not of the past. They had sensations but lacked the ability to compare them or form ideas. "The animals are therefore capable, like us, of both pain and pleasure; they do not recognize

2 For Buffon's attitude toward animals, see Duchet, *Anthropologie et Histoire*, 235-49; Robbins, *Elephant Slaves*, passim; Donald, *Picturing Animals*, 23, 33, 67, 103, 106, 165-6; Spary, *Utopia's Garden*, 99-117; and Hoquet, *Buffon*, 495-535.

3 "Les Animaux carnassiers," in Buffon, *Histoire naturelle*, 7: 3-38, at 4-9, 25-31.

4 Ibid., 3-4.

5 Tomaselli, "Moral Philosophy"; Whelan, "Population"; Blum, *Strength in Numbers*.

6 "Discours sur la nature des Animaux," in Buffon, *Histoire naturelle*, 4: 3-110, at 22-3.

good or evil, yet they feel them: that which is agreeable to them is good, that which is disagreeable is bad; the one and the other are only suitable or contrary connections with their nature, and their organization."[7]

Animals, like human beings, were unequal. Both among animals and among humans there was a scale of feeling ranging from automaton-like behavior to great sensitivity. Just as there was a difference in this respect between savages and civilized human beings, so a similar difference occurred between wild animals with natural sensibility and domestic animals with improved feelings.[8] This also had ethical consequences.

> The animals, at least those endowed with senses, flesh and blood, are sensible beings; like us they are capable of pleasure and subject to pain. It is therefore a type of cruel insensibility to unnecessarily sacrifice those especially who are similar to us, live with us, and who resemble us by displaying signs of pain; for those whose nature is different from ours, are hardly able to affect us.[9]

Natural pity was founded on an analogy with the suffering object, and therefore, while the suffering of a dog or a lamb excited human emotion, the destruction of a tree or an oyster did not. Buffon based his ethical outlook on the common eighteenth-century perception of "sympathy," which in modern terms means the ability to commiserate with another's feelings. Most commonly this applied to other human beings, but it could also apply to animals, if only to a limited extent. The common Enlightenment outlook exemplified so clearly by Buffon emphasized sensitivity to animal pain, but only so long as this did not collide with what seemed vital human interests, specifically as these related to the ongoing material and social progress of humanity. Despite his seeming interest in the traditional debate regarding the cognitive and sensual qualities of animals, Buffon was unwilling to let such concerns hamper his advocacy for the utilization of animals.

7 Ibid., 40-2: "Les animaux ont donc comme nous de la douleur & du plaisir; ils ne connoissent pas le bien & le mal, mais ils le sentent: ce qui leur est agréable est bon, ce qui leur est desagréable est mauvais; l'un & l'autre ne sont que des rapports convenables ou contraires à leur nature, à leur organization."

8 Buffon, "Les Animaux carnassiers," 8-9.

9 Ibid., 6-7: "Les animaux, du moins ceux qui ont des sens, de la chair & du sang, sont des êtres sensibles; comme nous ils sont capables de plaisir & sujets à la douleur. Il y a donc une espèce d'insensibilité cruelle à sacrifier, sans nécessité, ceux sur-tout qui nous approchent, qui vivent avec nous, & dont le sentiment se réfléchit vers nous en se marquant par les signes de la douleur; car ceux dont la nature est différente de la nôtre, ne peuvent guère nous affecter."

Buffon helps highlight a basic fact of Enlightenment considerations of animals: needless cruelty was condemned, but all too often the suffering of animals seemed lamentably necessary for human progress, in which case it was tacitly overlooked. The conjectural history so popular among Enlightenment intellectuals, not least in Scotland and France, required both moral improvement on the one hand, which in the case of animals meant ameliorating their treatment, and also, on the other hand, material progress, which meant increasing the scale and efficiency of the use of natural resources, including animals. When these two aims collided, it was the latter which needed to take precedence.

The tension between praise of the human command of nature on the one hand and sensitivity to animals on the other received one of its most eloquent eighteenth-century expressions in the writings of J. Hector St. John de Crèvecoeur. Combining the traditional praise of the moral worth of agricultural pursuits with a sense of the importance of commanding the American wilderness, Crèvecoeur extolled the human endeavor exemplified by the settlers of the New World. At the same time, however, he was enthralled by the awesome quality of nature in America, which offered such ample room for humanity's efforts to build a civilization on the basis of cultivation of natural resources. Animals, of course, were bound to be affected by this huge undertaking. Crèvecoeur, who became intimately aware of this influence, thus found himself expressing, probably unintentionally, the inherent Enlightenment tension between sensitivity to animals on the one hand and the ineluctable need to harm them in order to provide for human necessities on the other. In the case of the life of American farmers, these were indeed absolute necessities without which survival in the harsh wilderness became impossible.

In *Letters from an American Farmer*, Crèvecoeur's protagonist and mouthpiece James described his enjoyment of wild and domestic animals on his farm, combining appreciation of animals together with their utilization for human benefit.[10] He was astonished at the sagacity of animals which occasionally surpassed that of humans, and he objected to diminutive views regarding animal instinct.

> I never see an egg brought on my table but I feel penetrated with the wonderful change it would have undergone but for my gluttony; it might have been a gentle, useful hen leading her chicken with a care and vigilance which speaks shame to many women. A cock perhaps, arrayed with the

10 Crèvecoeur, *Letters*, 51-65.

most majestic plumes, tender to its mate, bold, courageous, endowed with an astonishing instinct, with thought, with memory, and every distinguishing characteristic of the reason of man.[11]

There was a remnant here of the old theriophilic comparison of humans with animals. But in contrast with earlier types of theriophilic literature, Crèvecoeur's recognition, both intellectual and also, as a settler of the wilderness, practical, of the need to utilize animals and other natural resources, made adherence to fine ethical principles quite difficult. In contrast with Montaigne, *The Turkish Spy*, or Rousseau, Crèvecoeur faced the ineluctable practical dilemma between sensitivity to animals and the need to exploit them for material and cultural progress. This dilemma had of course always existed for anyone sensitive to animal suffering, but it was only in the late Enlightenment that it began to be conceptualized in ever greater detail. Consequently, Crèvecoeur's protagonist James found himself in an ethical bind between his appreciation of animals and his constant need to utilize them. For example, he regretted that the bees on his farm were destroyed by kingbirds but also recognized that these birds guarded his fields against the depredations of crows. He therefore resisted killing them till they increased too much.[12] He continued and wrote of his enjoyment of quails and his attempts to help them during winter. "I do not know an instance in which the barbarity of man is so strongly delineated as in the catching and murthering those harmless birds, at that cruel season of the year."[13]

Both in *Letters from an American Farmer* and in *Sketches of Eighteenth-Century America* Crèvecoeur's struggle with this dilemma was constantly evident. He seemed to perceive a connection between political corruption and cruelty toward both animals and human beings.[14] As we saw in the case of John Oswald, this type of connection was still relatively rare even in the late eighteenth century. But in Crèvecoeur's case it was more significant, because his general outlook was much closer to the mainstream than that of the radical Oswald. At one point he wrote how, out of a "Pythagorean disposition," he refused to destroy an ant colony to satisfy his "impious" curiosity regarding its structure. But at another time he destroyed ants which threatened a beehive, and Pythagorean vegetarian allusions

11 Ibid., 55. On animal sagacity and other animal qualities, see also ibid., 61-2, 120, 305-7.

12 Ibid., 55-6.

13 Ibid., 56-7.

14 This seems to be the implication of the passages in ibid., 418-19.

notwithstanding, when noting the consumption of snapping turtles, he claimed that "as long as we feed on what would feed on us, that seems to be founded on a just retaliation."[15]

Other animals, however, seemed harmless, and therefore were true victims of unjustified human barbarity. Beavers, of course, were one of the most ubiquitous examples of hunted animals. They were "the philosophers of the animals; the gentlest, the most humble, the most harmless. Yet brutal Man kills them. I was once a witness to the destruction of one of their associated confederacies. I saw many of them shed tears, and I wept also; nor am I ashamed to confess it."[16] Nevertheless, the need to kill animals seemed at times a truly lamentable necessity. Regarding the regrettable need to hunt squirrels which harmed the crops, Crèvecoeur noted that "For my part, I cannot blame them [the squirrels carrying away corn husks], but I should blame myself were I peaceably to look on and let them carry all. As we pay no tithes in this country, I think we should be a little more generous than we are to the brute creation." Even when one could not avoid harming animals, this had to be done while making an effort to mitigate their plight. This was evident in the battle of farmers against birds which harmed their crops.

> But after all the effects of our selfishness, are they [the blackbirds] not the children of the great Creator as well as we? They are entitled to live and to get their food wherever they can get it ... But Man is a huge monster who devours everything and will suffer nothing to live in peace in his neighbourhood ... Their [the blackbirds'] depredations proceeded from hunger, not from premeditated malice ... it is necessary that either Man or Nature should feed them.[17]

Nevertheless, Crèvecoeur did not at any point deny the evident necessity to make constant use of animals and to destroy them whenever the need arose. Hunting animals did not necessarily involve inherent ethical qualms. Crèvecoeur's positive depiction of whaling in the Cape Cod islands was a case in point.[18] To modern sensitivities this might seem particularly blameworthy, although subsequent recognition of the mental capacities

15 See the respective remarks in ibid., 244-9, 295, 301, and also see 398, and for further appreciation of bees, 55-6, 315.

16 Ibid., 301.

17 Ibid., 292-4.

18 Ibid., 163-4.

of cetaceans was still not evident in the eighteenth century. This was not the case, however, regarding other animals. Thus, regret at the suffering of blackbirds notwithstanding, there was no hint of criticism of cruelty in the depiction of the hunting of pigeons with a net, "to which they are allured by what we call tame wild pigeons, made blind and fastened to a long string."[19] This seeming inconsistency of Crèvecoeur's should come as no surprise, however. It was perfectly in tune with the prevalent attitude of his contemporaries toward animals. Crèvecoeur's writing mixed literary musings together with realistic depictions of frontier life, the latter imbued with the Enlightenment praise of cultural progress. The literary aspect accommodated poignant sensitivity to animals, yet the overall realistic underpinning of the discussion left much less room for this ethical concern.

The Enlightenment outlook often conceived of mitigating the suffering of animals as a requirement for improving the world in which human beings lived. It was part of the progress of civilization in general. Modern humanitarianism, as it began emerging during the eighteenth century, was from the start primarily concerned with ameliorating the human condition. When animal welfare conflicted with this overriding aim, the result was preordained. As we have already noted, it was precisely the increasing remove of urban populations from direct contact with nature and wild animals which enabled a rising sentimental concern with animal feelings. Crèvecoeur was an exception to this rule. Buffon, with his scientific investigations conducted in the safety of the Parisian Jardin du Roi, was much more representative of most of the authors who wrote about animals in the eighteenth century, those composing travel books excepted. Yet both Crèvecoeur and Buffon exemplified the same dilemma, a dilemma which was increasingly accentuated with the rise of modern industrial production and its consequent cultural, social, and economic influences.

Buffon and Crèvecoeur also pointed the way toward the real innovation of Enlightenment thought regarding animals: their consideration as resources to be used for the progress of human civilization. Moral complications aside, human progress was considered as the ultimate aim which justified the harm and suffering which regrettably often accompanied historical development. The remainder of our discussion will focus on this shift toward a historical and economic consideration of animals.

19 Ibid., 61. For uncritical mentions of hunting, see also ibid., 92-3, 286. For more regarding Crèvecoeur's views on hunting, see Allen and Asselineau, *St. John de Crèvecoeur*, 41-5. On his general attitude toward animals, see also Philbrick, *St. John de Crèvecoeur*, 97-106.

8 Animals in Enlightenment Historical Literature

Buffon and Crèvecoeur demonstrated the influence of the new Enlightenment conjectural theorizing about the history of human civilization. Yet they were not historians. The eighteenth century saw an unprecedented interest in both traditional military and political historical topics and new cultural and social ones. This was accompanied by a new reflexive historical awareness. Enlightenment intellectuals developed a new conception of their own culture and how it differed from earlier eras. Norbert Elias, in his famous book *The Civilizing Process*, viewed the late eighteenth century as precisely the time when the civilizing process became manifest in western civilization, when human beings became consciously aware of their actually being "civilized." It was only then that the concept of "civilization," as a new bourgeois reflexive term, itself appeared.[1] This entailed a distinctly *historical* awareness that humanity was undertaking a long historical mission of advancement toward ever growing progress. As we have already begun to see, this new awareness accorded an important place to the human command of nature, including animals, for the sake of such progress. Important aspects of modern attitudes toward nature and animals crystallized into their more recognizable contemporary forms during the late Enlightenment. Of course the Darwinian Revolution later added a further vital component to the evaluation of nature, and humanity's place in it. But its influence in this respect was built upon the foundations of the preceding century, enhancing rather than replacing them. While an evolutionary perspective could be utilized to advance sensitivity to animals, it could also serve to justify their subjugation, as indeed the subjection of human beings by each other. The democratic implications of Darwinism were theoretically obvious, but their practical realization was more difficult.[2]

The emergence of the modern conceptualization of animals, and their place in human civilization, was a development of the second half of the eighteenth century. The fact that the same period also saw the rise of an unprecedented historical self-awareness was anything but fortuitous. It is therefore no surprise that late eighteenth-century historical literature,

1 Elias, *Civilizing Process*, 3-41.

2 For different interpretations of Darwin's influence on the debate about animals, compare Rachels, *Created from Animals*; Preece, "Darwinism."

in an unprecedented manner, began addressing the importance of the human use of animals as an essential part of the tale of human progress. The fact that animals received growing attention from historians was in large measure also owing to the Enlightenment ideal of *l'historien philosophe*. Eighteenth-century historiography consisted mainly of three components – erudition, narrative, and philosophy.[3] The latter was most typical of the times, referring to the expectation from historians to address the Enlightenment battle with superstition and despotism, but also opening the door for more modern subjects such as social and cultural history.[4] It was therefore no surprise that the animal theme eventually received attention from historians. In what follows we shall concentrate particularly on the works of Herder, Raynal, Robertson, and Gibbon, although we shall also mention other eighteenth-century historians.

Late Enlightenment historiographical considerations of animals were significantly influenced by the emergence of new forms of discussion of the origins of human civilization, most notably conjectural history, in effect the beginning of what would later become modern theoretical anthropology. Buffon's role in this intellectual venture has already been discussed. One cannot exaggerate the impact that his discussion of the human command of nature had on his contemporaries.[5] It presented them with a clear depiction of how mastery of nature, including of course animals, fueled the ongoing progress of civilization. Rousseau's two famous discourses were examples of this type of discussion. Rousseau, however, was untypical for his times, as he used conjectural history to undermine the worth of unhampered cultural progress. The outlook of most other Enlightenment intellectuals was quite different. It was the Scottish Enlightenment in particular which made conjectural history a truly rigorous intellectual tradition, in the process praising cultural progress. The most influential variation of Scottish conjectural history was stadial theory, which perceived various stages in the development of human civilization. The number of stages varied, but the most popular version was the four-stages one, propounded in one of its clearest versions by Adam Smith.[6] The four-stage theory outlined how human societies advanced according to their modes of subsistence, from a

3 Pocock, *Narratives of Civil Government*; Momigliano, "Gibbon's Contribution."

4 Phillips, *Society and Sentiment*. On Enlightenment historiography in general, see e.g. O'Brien, *Narratives of Enlightenment*.

5 On Buffon's influence, see Reill, "Buffon and Historical Thought." For the Scottish Enlightenment scene, see Wood, "Buffon's Reception."

6 Meek, *Social Science*; Pocock, *Barbarians, Savages and Empires*; Hont, "Language"; Berry, *Social Theory*, 93-9. Adam Smith's most clear exposition of the four-stage theory can be found

hunting stage, through a shepherding, then an agricultural stage, and finally reaching the stage of advanced commercial society. The importance for stadial theory of the need for commanding nature was clearly evident. Each of the stages added new forms of cultivating nature and utilizing animals. Even the commercial stage, which the Scottish philosophers considered as the most advanced form of social and economic civilization, included many uses of animals and animal products.

As innovative as stadial theorizing was in its new form in the second half of the eighteenth century, it was in fact built on a long tradition of historical analysis, which may help explain its significant influence on the formation of modern historical and social thought. Like many important intellectual innovations, it owed its impact not just to its novel ideas but also to its reformulation of notions which had been part of the conceptual landscape for centuries. It should be kept in mind how immersed eighteenth-century intellectuals were in classical history and scholarship, where they encountered early stadial notions. Thucydides was already aware of the advantages which sedentary civilization had over nomadic existence for the development of agriculture and commerce.[7] This differentiation between vagrant and sedentary existence would become key in Enlightenment historiography. Later, Varro was among the classical authors who made even more explicit stadial observations, utilizing stadial terminology in describing the rise of material civilization.[8] Particularly important in this regard was Ammianus Marcellinus, not least because of his influence on Edward Gibbon.[9] Ammianus's stadial observations were highly developed and unprecedentedly detailed, and he particularly recognized the need for a sedentary civilization as a precondition for advancing agriculture.[10]

Such stadial notions were less common in medieval historiography. But these classical examples were well familiar to eighteenth-century scholars when they became increasingly more prone to examining historical development through a stadial prism. Giambattista Vico noted that the order of human institutions was "first forests, then huts, next villages, later cities, and finally academies."[11] Like much of his theorizing, Vico's stadial scheme

in his *Lectures on Jurisprudence*, 14-16. Also see Pocock, *Narratives of Civil Government*, 309-29; Pitts, *Turn to Empire*, 25-58; Marouby, "Adam Smith."

7 Thucydides, *History*, 35-6 (I.2).

8 Cato and Varro, *Agriculture*, 313-15 (Varro, II.i.3-5), and also 175-7 (I.ii.12-16), 423-7 (III.i.1-8).

9 Womersley, *Transformation*, 169-81.

10 Ammianus Marcellinus, [*Works*], 3: 385 (XXXI.2.10), and also 1: 27 (XIV.4.3), 2: 237 (XXII.8.42), and passim.

11 Vico, *New Science*, 98-9.

was idiosyncratic. He did not emphasize the significance of commerce, and he placed the development of agriculture ahead of herding.[12] Yet he consistently recognized the stadial aspect of historical development and the significance that cultivating nature had in this process.[13] Vico was not influential in the eighteenth century, however. The same, of course, could not be said of Montesquieu, who in his own way also evinced a rather critical recognition of the stadial aspect of historical progress and the significance of agriculture, viewing commerce and artisanship as inimical to Roman culture in its finest moments, which were based more on military virtue.[14] Yet this did not invalidate the fact that historical development was indeed stadial by nature. Montesquieu expressly differentiated between savages and barbarians, claiming that the former were usually scattered people subsisting by hunting while the latter were pastoral nations capable of uniting together.[15] While the stadial concepts of Vico and Montesquieu were not as developed as those of the second half of the eighteenth century, they did set the stage for the noticeably new detailed elaboration of the stadial historiographical outlook which was shortly to emerge. Together with Buffon's theories, stadial theorizing was to constitute a firm intellectual underpinning for the late Enlightenment commendation of utilizing nature and animals for human progress.[16] Historians in particular incorporated the insights offered by these influential new considerations of the rise of civilization through the mastery of nature. By the late eighteenth century it became virtually impossible to consider the history of human civilization without noting the importance of the command of natural resources as a precondition for higher cultural development. In this light it was all but inevitable that historians would come to consider the utilization of animals as part of this broad historical process.

Not all Enlightenment historians, however, contributed to this new consideration of animals. A case in point regarding the uniqueness of the historiographical outlook on animals was David Hume, who as a philosopher was quite interested in discussing animals but in his famous *History of England* veritably ignored this issue.[17] Voltaire was only slightly

12 Ibid., 15.

13 Ibid., 11-12, 308-9, 470.

14 Montesquieu, *Considerations*, 27, 98-9, 137, 164-5.

15 Montesquieu, *Spirit of the Laws*, 290-1.

16 On nature in eighteenth-century conceptions of progress, see Spadafora, *Idea of Progress*, 56-8, 284, 287-8.

17 The closest he came to addressing this issue in his historical work was in Hume, *History of England*, 6: 142.

more prone to evince his ethical sensitivity to animals in his historical rather than his philosophical works.[18] Two reasons might explain why these two prominent figures in the development of Enlightenment historiography veritably ignored animals in their historical works. First, they may have considered that they had devoted sufficient attention to this topic in their philosophical writings. But second, and more important, they were both less influenced by stadial theory than most other late Enlightenment historians. This is more surprising regarding Hume than the somewhat older Voltaire. Hume, after all, played a vital role in the Scottish Enlightenment emphasis on political economy and economic history.[19] Yet in this sense, at least, he was an exception to the rule. If anything, this emphasizes even more the general connection between stadial theory, and conjectural history in general, and the rising late Enlightenment emphasis on the importance of utilizing animals. Other contemporary historians consequently demonstrated a consistent interest in animals unprecedented in earlier historical literature.

Johann Gottfried Herder's views on animals, specifically in *Outlines of a Philosophy of the History of Man*, have received little scholarly attention.[20] More than any other Enlightenment historian (and here he should be considered primarily as a historian), his discussions of animals combined philosophical discourse with historical comments. Herder thus offers the opportunity of assessing how the Enlightenment historiographical outlook on animals was both influenced by, and yet departed from, more traditional debates about them. Although most of his French and British contemporaries, particularly among prominent historians, seem to have been unfamiliar with his work, he himself was quite familiar with contemporaneous intellectual discourses in other countries, such as Raynal's work and in particular the works of the Scottish Enlightenment, including those of William Robertson. It is therefore not surprising that he was also influenced by stadial theory.[21]

18 See the mainly cursory treatments in Voltaire, *Essai sur les mœurs*, 1: 23-5, 60-1, 247-8, 818; 2: 314.

19 Wootton, "David Hume."

20 Although see Steiner, *Anthropocentrism and Its Discontents*, 180-4. Also see Muthu, *Enlightenment Against Empire*, 232-4, 240-1. For Herder's general cosmological ideas, see Glacken, *Traces on the Rhodian Shore*, 537-43, 569-70; and Clark, *Herder*, 308-47.

21 For Herder's familiarity with Raynal's work, see Pagden, *Lords of All the World*, 162-3; and, particularly regarding Herder's familiarity with Diderot's contributions to Raynal's work, see Muthu, *Enlightenment Against Empire*, 72, 121, 211. For Herder's cautious use of stadial theory, see ibid, 238-46, and for his familiarity with William Robertson's works, ibid., 217. For the general influence of Scottish Enlightenment ideas on Herder, see Oz-Salzberger, *Translating the Enlightenment*, 19-20, 22, 30, 34, 57, 65, 70-3, 103, 130, 135, 152, 184, 315, 318.

Yet his approach to the topic of animals was less original than that of the other eighteenth-century historians we will be examining, and, in particular, he was less emphatic when it came to assessing the material utilization of animals. Herder, aiming at the now-defunct Cartesian beast-machine theory, claimed that animals were not machines and that to regard them as such was "a sin against Nature."[22] Every creature had a place in the world.[23] Human mental superiority was more a matter of degree than kind. For example, all highly developed animals loved their offspring. This love was greater the more a creature was advanced, and therefore most conspicuous in human beings. Consequently, maternal love was the basis for the inborn sociability of humanity.[24] Even animals obeyed the laws of justice, and human beings who did not do so were "the most inhuman of all creatures, even if they be the kings and monarchs of the Earth."[25] Only human beings, however, had speech, which made human cultural development possible.[26] Also unique to humanity was the capacity for religion, "the highest humanity of mankind."[27] The ability to walk on two legs was particularly important, enabling the use of free hands and even speech.[28] Presenting his version of a common early modern argument for human superiority, Herder noted that humans did not have any one characteristic in perfection like other creatures, but a combination of a certain level of all their characteristics.[29] Rationality and the capacity for liberty were what made human beings unique.[30] Other animals did not harm each other like human beings, yet only humans had speech, traditions, religion, and laws. The manner in which humanity acquired its uniqueness was a historical issue, connected to the peculiar human capacity for perfectibility or corruptibility.[31] Herder noted the possibility of human ethical obligations in treating animals.[32] Human beings became wild as a result of contact with the blood of animals, hunting, war, and other human mischiefs.[33] Human moral capacity was superior to that of animals, and human compassion was able to commiserate with the

22 Herder, *Outlines*, 66. The original work was published in 1784-91.
23 Ibid., 50.
24 Ibid., 48, 101.
25 Ibid., 102.
26 Ibid., 43, 233-7, 240.
27 Ibid., 103-5, 251-6.
28 Ibid., 69-89, 95-8, and passim.
29 Ibid., 40-1.
30 Ibid., 89-94, and also 229-30, 282.
31 Ibid., 67.
32 Ibid., 34.
33 Ibid., 282. And see also the remarks on metempsychosis at 304, 309.

feelings of other creatures, even worms and plants. Herder conclusively inferred from this a condemnation of animal experimentation. "He must possess rigid nerves, who can open a living creature and watch its convulsive movements: nothing but an insatiate thirst for fame and science can gradually deaden his organic sensibility."[34]

Most of the above remarks were old-fashioned, in the tradition of moderate theriophily. Nevertheless, there were clear limits to Herder's sensitivity to animals. He propounded an idiosyncratic version of the theory of the Great Chain of Being.[35] He claimed that the more one rose in the scale of living creatures, the more one's ability to adapt to different states was augmented. "Of all these changeable, modifiable, adaptable creatures, man is the most adaptable: the whole Earth is made for him; he for the whole Earth."[36] "Adapt" here did not mean simple accommodation with natural conditions. On the contrary, it signified an active human endeavor to overcome these natural limitations. What was adapted was the natural environment to human beings, not the other way around. Human superiority stemmed above all from the distinctly *historical* process of gradually commanding nature. Here Herder was closer to the more original historiographical emphasis on animals as instruments for human progress.

Herder emphasized the plenitude of creation, the view that in nature everything that might exist did in fact exist.[37] There was a specifically *Sturm und Drang* dialectical twist to this perception, however. Nature endeavored to create the maximum number of creatures peacefully coexisting, yet this necessitated a violent struggle between them, which led to an equilibrium of powers as a prerequisite of such peace. Human struggle with other creatures was therefore part of the natural order.[38] Human dominion was not absolute and was based on a constant battle with animals.[39] Like most philosophers inclined to a benign view of animals, Herder retained an overall anthropocentric outlook, asserting that without humanity the

34 Ibid., 99-100.

35 Herder's version of this theory is outlined mainly in ibid., 107-31. He utilized it as a veritable proof for the possibility of human immortality as a higher form of existence. For the claim that Herder did not in fact agree with the Great Chain of Being theory, see Reill, *Vitalizing Nature*, 186-91. Reill claims that Buffon also disagreed with the Great Chain of Being theory. See ibid., 49-50, 164.

36 Herder, *Outlines*, 11.

37 Furthermore, human beings were superior because they could understand the language of creation and were therefore the images of God. See ibid., 463-5.

38 Ibid., 35-8.

39 Ibid., 35.

earth would lack its "supreme ornament and crown."[40] Human dominion over the earth was "the most ancient philosophy of the history of man."[41] Yet human superiority was not just a result of a preordained religiously based cosmological order; it was also a historical construction, resulting from the developmental nature of human beings.[42] It was on this point that Herder's view of animals demonstrated the novel outlook of late Enlightenment historical literature. Most of his other claims regarding animals were stock items in the early modern philosophical repertoire. Yet Herder also considered the place that utilization of nature played in the progress of human civilization. Propounding a version of stadial theory, he claimed that each area and period of the earth enabled a different state of human culture. All of these, considering the plenitude of nature, unavoidably existed.[43] Geography directly influenced human nature in various regions.[44] The most perfectly formed humanity was in the Mediterranean, specifically ancient Greek culture, which later constituted the basis for European supremacy.[45] Probably influenced by Buffon, Herder claimed that human beings could control nature, however, and could even improve the climate, in a process which might continue to an unforeseeable extent.[46] Retaining traditional religious notions of human superiority, Herder, in early romantic fashion, also emphasized the essentially violent underpinning of the human interaction with nature.[47] The novel aspects of Herder's discussions of animals were still predominantly couched within the traditional debate about animal feelings and rationality. Herder's views thus form a conceptual bridge. On the other side of this bridge were situated some of the most prominent historians of the late Enlightenment, who combined their considerations of animals within their broader historical analyses in a much more pronounced manner than he did.

This was amply evident in the Abbé Raynal's great historical work on the European colonies, one of the most widely read works of the late eighteenth

40 Ibid., 70.
41 Ibid., 278.
42 Ibid., 129.
43 Ibid., 18-19.
44 Ibid., 35-8, 173-4, and passim.
45 Ibid., 144-5. Herder was more of a European chauvinist than a racist. See especially the remarks in ibid., 132-62. This of course did not preclude the abuses to which his philosophy later gave rise. The important and unresolved contentious debate regarding Herder's possible Counter-Enlightenment outlook lies outside the scope of our discussion, though in my view he was predominantly an Enlightenment philosopher.
46 Ibid., 176.
47 Compare the remarks in Ferguson, *Essay*, 7, 10-11, 16-17, 35-6, 48-50, 207.

century, which also included contributions by other authors, most notably Denis Diderot.[48] Raynal claimed that human beings' cruelty toward each other demonstrated how they were the most wicked and unfortunate of animals.[49] He was emphatically in favor of ethical sensitivity to animals, and the most theriophilic discussion in all of eighteenth-century historical literature was his depiction of the effects of widespread hunting on the animals of America.[50] In this context the beaver received the most extensive treatment. "[T]his harmless animal, who never hurts any living creature, neither carnivorous nor sanguinary, is become the object of man's most earnest pursuit, and the one which the savages hunt after with the greatest eagerness and cruelty: a circumstance owing to the unmerciful rapaciousness of the most polished nations of Europe."[51] The beavers demonstrated sociability, gentleness, and particularly intelligence, noted by their ability to build dams, and Raynal also depicted their monogamous love life and their parental cares.[52] The beaver was "Happy, if his coat did not tempt merciless and savage man to destroy his buildings and his race." Despite the superiority of human mental faculties, the beaver proved itself superior in social attainments.[53] Man's superiority was simply due to his free hands. He "marks his dominion … over the earth, by destroying and ravaging the face of the globe."[54]

Nevertheless, these assertions were conspicuous precisely because they were so untypical for Raynal. Much more than Herder, his outlook exemplified the essential difference between traditional philosophical discussions of animals and the new, distinctly historiographical appreciation of their importance. Perhaps these inconsistencies were a result of the multi-authorship of his work. However, from a more dialectical perspective, this inconsistency can be interpreted as an inherent tension within the rising modern consideration of animals, attempting to accommodate sensitivity to animals together with a growing advocacy of the cultural need to utilize

48 Raynal, *Philosophical and Political History*. For the sake of simplicity, in what follows I shall disregard the complicated implications of the plural authorship of this work, although we should note that Diderot is credited with most of the more radically democratic passages in the later editions.

49 Ibid., 6: 140-1, and also 149.

50 Ibid., 5: 150, 163-4.

51 Ibid., 5: 164-77, quotation at 167.

52 Ibid., 5: 167-72.

53 Ibid., 5: 173-5.

54 Ibid., 5: 175-6. Elsewhere, however, at 3: 155, he claimed that reason was the only advantage human beings had over other animals. Regarding the beaver, see the discussion of Raynal's attitude toward nature in Womack, "Eighteenth-Century Themes," 249-59.

them. Just like Raynal's praises of savage societies, so his sensitivity to animals was limited in scope. Ultimately, he regarded advanced human civilization in its European form as the best possible course for human progress. Cruelty toward animals required amelioration, yet using them remained vital for human advancement. Similarly, Raynal wished to improve the treatment of primitive peoples, thus enabling them to profit from their contact with advanced Europeans without being exploited. Therefore, his discussion of the beaver notwithstanding, Raynal's customary approach to animals was quite different. He claimed that carnivorous human beings were more robust and healthy than vegetarians. Man was able to accommodate himself to every type of climate. "[H]e therefore feeds upon the spoils of the chase, upon the produce of the waters, or upon the fruits of the earth; and becomes either a shepherd or a husbandman, according to the fertility or barrenness of the soil he inhabits."[55] Raynal noted cases of kindness to animals but condemned despots who put the welfare of their pets above that of their human subjects.[56] He criticized the theory of the Great Chain of Being, asserting that there was a great interval between superior man and all other animals.[57] In contrast to the remarks on the beaver, he was usually totally uncritical of hunting.[58] He associated hunting with courage and a martial spirit, and admittedly also with cruelty, but mainly toward other human beings.[59]

Most significantly, Raynal was constantly aware of the importance of the utilization of animals for the progress of civilization. For instance, he noted the importance of the camel and horse for Arabian culture, as of the lama for South American cultures.[60] When human beings failed to command nature, both they and nature itself deteriorated. Voicing a common early modern observation about the degeneracy of America compared to Europe, Raynal noted the inferior quadruped animals in the New World, many initially imported from Europe, which would not have degenerated given sufficient attention by the settlers.[61] Any notion of the preferment of pristine

55 Raynal, *Philosophical and Political History*, 1: 72-3.

56 Ibid., 2: 31, 46-7; 5: 32-3, 126. See also the critical remarks on metempsychosis at 1: 51, 66-7; 2: 325, and the remark on kindness to animals in India at 6: 478.

57 Ibid., 3: 352-3.

58 Ibid., 3: 170-1. See similarly the remarks on the hunting and use of seals at 5: 251-2.

59 Ibid., 5: 113, 141, 149.

60 Ibid., 1: 333-5; 3: 106-9.

61 Ibid., 4: 132-5; and see also the remarks at 6: 87. On the heated Enlightenment debate about the purported degenerative qualities of America, see Gerbi, *Dispute of the New World*; and Chinard, "Eighteenth Century Theories."

as opposed to cultivated nature had no room in his quintessential Enlightenment outlook. Animals, like nature in general, were meant for human use. While Raynal based his sanction of human mastery of nature on rational scientific argumentation, he was, like many other early modern intellectuals, transmuting, rather than replacing, the religious anthropocentric cosmology, thus creating an even more potent sanction for the utilization of nature. It was this combination of the old and the new which gave rise to the novel and unprecedented modern emphasis on commanding nature. Raynal's emphatic advocacy of cultivating nature was evident when he depicted the desolate state of North American nature before the arrival of the European settlers. However, then

> man appeared, and immediately changed the face of North America. He introduced symmetry by the assistance of all the instruments of art. The impenetrable woods were instantly cleared, and made room for commodious dwellings. The wild beasts were driven away, and flocks of domestic animals supplied their place; while thorns and briars made way for rich harvests. The waters forsook part of their domain, and were drained off into the interior parts of the land, or into the sea by deep canals. The coasts were covered with towns, and the bays with ships; and thus the New World, like the Old, became subject to man.[62]

Clearly evincing Buffon's influence, Raynal here outlined the typical late Enlightenment advocacy for the human mastery of nature and animals as an essential prerequisite for cultural progress. A similar consideration of animals was common in the historical literature of the second half of the eighteenth-century. The manner in which they were treated seemed important only to the extent that it furthered human interests. The French historian Antoine-Yves Goguet, for example, noted as typical of the Athenians' kindness their benign handling of laboring animals.[63] Goguet, however, shared the traditional notion of human superiority to animals, evinced in large measure by the capacity for speech.[64] Even when he noted sensitivity to animals, this was based on the traditional assertion that their treatment influenced the treatment of human beings. Thus Goguet claimed that ancient legislators ruled over ferocious people who had just emerged from barbarism, and he had "no doubt therefore, but those legislators had a view

62 Raynal, *Philosophical and Political History*, 5: 302-3.

63 Goguet, *Origin of Laws*, 3: 232.

64 Ibid., 1: 71.

of inspiring their people with gentleness and compassion towards each other, by inspiring them with these sentiments towards animals."[65]

Not surprisingly, a similar outlook was demonstrated by Goguet's English translator, the Scottish historian Robert Henry, when he noted the ancient British laws protecting domestic animals, claiming they were motivated by economic considerations.[66] The history of animal domestication afforded amelioration of human life. In Roman Britain dogs were useful companions, but only at a higher cultural stage did they also serve for entertainment.[67] Henry discussed them primarily through their uses for humans. He was uncritical of hunting, treating it as a preparation for war in a romantic vein based on the Ossianic poems.[68] He observed how the tenth-century king Edgar had imposed an uncommon form of tribute on the Welsh princes – in place of money and cattle as formerly, requiring instead three hundred wolves' heads annually. This resulted in keen hunting of "these destructive animals," whose numbers rapidly declined in a few years.[69] Henry did note William the Conqueror's "horrid cruelties" in his excessive passion for hunting, but whether he meant cruelty toward animals or toward human beings was unclear.[70] More typically he mentioned animal baiting without noting its cruelty, and when he did note the cruelty involved in hunting, it was toward socially inferior human beings, not toward animals.[71]

Like most Scottish Enlightenment intellectuals, Hume notwithstanding, Henry was influenced by stadial theory.[72] The same could be said regarding the most prominent contemporary Scottish historian (with the possible exception of Hume), William Robertson. Robertson was particularly interested in how utilizing animals furthered cultural progress, exhibiting almost no interest in the traditional debate about animal sentient and cognitive qualities. Noting the beneficial uses of the horses that the Spaniards under Columbus had brought to the New World, he wrote of "that vast accession

65 Ibid., 1: 35-6. For Goguet's stadial notions, see 1: 84-5, 277. On Goguet in general, with further references, see Wolloch, "'Facts, or Conjectures'."

66 Henry, *History of Great Britain*, 1: 215-16. Henry's responsibility for translating Goguet is commonly accepted, but not certain.

67 Ibid., 1: 390-1.

68 Ibid., 1: 487-8.

69 Ibid., 2: 76. Henry's positive view of the extirpation by hunting of "these destructive animals" was clear also at 5: 563.

70 Ibid., 3: 27.

71 Ibid., 3: 595-6.

72 For Henry's historiography in general, with further references, see Wolloch, "History 'On a New Plan'."

of power, which man hath acquired by subjecting them to his dominion."[73] This subjection of animals was one of the most important foundations of human progress.

> As the people of the East appear soon to have acquired complete dominion over the useful animals, they could early undertake the long and toilsome journies [*sic*] which it was necessary to make, in order to maintain this intercourse [between different countries]; and by the provident bounty of Heaven, they were furnished with a beast of burden [the camel], without whose aid it would have been impossible to accomplish them.[74]

Typically for the Scottish Enlightenment, Robertson stressed the importance of commerce, the final stage in the four-stages scheme. Therefore the camel's furthering of trade added to its significance for cultural progress. More generally, Robertson noted how

> The greatest operations of man in changing and improving the face of nature, as well as his most considerable efforts in cultivating the earth, are accomplished by means of the aid which he receives from the animals whom he has tamed and employs in labour ... But man, in his civilized state, is so accustomed to the service of the domestic animals, that he seldom reflects upon the vast benefits which he derives from it. If we were to suppose him, even when most improved, to be deprived of their useful ministry, his empire over nature must in some measure cease, and he would remain a feeble animal, at a loss how to subsist, and incapable of attempting such arduous undertakings as their assistance enables him to execute with ease.[75]

Robertson's recognition of the importance of animals for human progress was repeatedly expressed in his denigration of the indigenous Americans.[76] He claimed that they were insensible both to animal suffering and to that

73 Robertson, *History of America*, 1: 171.

74 Robertson, *Historical Disquisition*, 3. For more on the importance of the camel for transportation and trade in desert regions, see ibid., 400-2.

75 Robertson, *History of America*, 2: 124-5.

76 Robertson's and other contemporaries' denigration of America and its inhabitants was countered by the Mexican-born Jesuit historian Francisco Clavigero. Clavigero, however, was not particularly interested in animals, making only casual remarks about them in a rather anthropocentric vein. See his *History of Mexico*, 1: 48, 242-3, 380-1. For Clavigero and his criticism of Robertson, see Pocock, *Barbarians, Savages and Empires*, 205-26.

of other human beings, including their own family members. They treated domestic animals harshly, "never employ[ing] any method either for breaking or managing them, but force and cruelty."[77] It should be noted that Robertson was not the only one in the eighteenth century who presented a picture of primitive cruelty which runs counter to modern notions of primitive innocence.[78] Primitivism, perhaps even more than sensitivity to animal sentience, was a minority outlook.

More important for Robertson, however, were the specifically material-cultural advantages which the command of animals offered. This was conspicuously lacking among the South Americans, as was their general want of cultivation of nature. The indigenous Americans, including the advanced civilizations of Mexico and Peru, lacked the technological utilization of metals and the control of animals for labor and agriculture.[79] One consequence of the physical inferiority of America and its uncultivated condition was a scarcity of animals, which were also inferior in size and robustness compared with those of other continents. What did exist in America was an unusual proliferation of various noxious animals, specifically because of lack of cultivation.[80] Robertson claimed that America originally did not have domesticated animals and that therefore its peoples developed in isolation from each other. "Whenever any people have experienced the advantages which men enjoy, by their dominion over the inferior animals, they can neither subsist without the nourishment which these afford, nor carry on any considerable operation independent of their ministry and labour."[81] The use of animals was an important indicator of cultural progress, yet the people of America, in contrast even with the Tartars, had not acquired it.[82] "This command over the inferior creatures is one of the noblest prerogatives of man, and among the greatest efforts of his wisdom and power. Without this, his dominion is incomplete. He is a monarch, who has no subjects; a master, without servants, and must perform every operation by the strength of his own arm."[83] Even the advanced Mexican and Peruvian cultures, with

77 Robertson, *History of America*, 2: 216-17.

78 See e.g. Adair, *History of the American Indians*, 405. Adair claimed that North American Indians killed too many buffalo in a wasteful manner, leaving most of the carcass and keeping only the tongue and marrow-bones. For criticism of the modern notion of the innocence of primitive people's treatment of animals, see Preece, *Animals and Nature*.

79 Robertson, *Historical Disquisition*, 214.

80 Robertson, *History of America*, 2: 18-23.

81 Ibid., 2: 35-6.

82 Ibid., 2: 122-5.

83 Ibid., 2: 122-3.

their use of domesticated animals, did not shake Robertson's views.[84] He claimed this domestication was only of a basic kind and applied mainly to the llama, though even it was not sufficiently utilized. This was a prime reason why both these nations were only in the infant stage of human progress upon the arrival of the Spaniards.[85]

The views on the cultural importance of commanding nature and animals expressed by the greatest of Enlightenment historians, Edward Gibbon, displayed a similar outlook. As with so many other topics, his work, specifically *The History of the Decline and Fall of the Roman Empire*, presented a *summa* of Enlightenment historiography's philosophical outlook. Gibbon was highly influenced by Buffon and by stadial theory.[86] The importance of the human command of nature and animals for cultural progress was repeatedly evident in the *Decline and Fall*.[87] Though he vacillated regarding the medical influences of a carnivorous diet and the possible connection between cruelty and meat eating, Gibbon regarded the Rousseauist advocacy of vegetarianism as more emotional than rational.[88] He was much closer on this issue to Buffon, who opposed unnecessary harm to animals but regarded consumption of animal flesh as part of the natural order.[89] If there was a connection between cruelty and a carnivorous diet in Gibbon's view, it was most emphatic in barbaric Tartar society, where animal slaughter was openly done by the "unfeeling murderer," while "the horrid objects" [animal carcasses] were "exhibited in their naked and most disgusting simplicity, in the tent of a Tartarian shepherd," but also "disguised by the arts of European refinement."[90]

84 Ibid., 3: 12.

85 Ibid., 3: 152-4; and see also 3: 217-18. At 3: 311-12, Robertson noted the rapid proliferation of domesticated European animals once they were brought to America. For Robertson's criticism of the Mexicans and Peruvians, see Hargraves, "Beyond the Savage Character."

86 As J. G. A. Pocock has observed: "A door to the future was opened, in Gibbon's mind, by Buffon's decision to treat the human as an animal species like any other." See Pocock, *Narratives of Civil Government*, 362; and also Pocock, *Barbarians, Savages and Empires*, 44. On the influence of stadial theory, see Pocock, "Gibbon and the Shepherds." For examples of Gibbon's stadial observations, see Gibbon, *History of the Decline and Fall*, 1: 238-9, 996-9, 1027-9; 2: 819-20; 3: 156, 449.

87 On the importance of cultivating nature, see particularly the remarks in "General Observations on the Fall of the Roman Empire in the West," in Gibbon, *History of the Decline and Fall*, 2: 508-16. Also see Wolloch, *History and Nature in the Enlightenment*; and Wolloch, "Edward Gibbon's Cosmology."

88 Gibbon, *History of the Decline and Fall*, 1: 1026-7, esp. 1026, note 9, where Gibbon quotes Rousseau's *Emile* and also mentions Plutarch and Ovid in relation to vegetarianism.

89 "Les Animaux carnassiers," in Buffon, *Histoire Naturelle*, 7: 4-9, 25-31.

90 Gibbon, *History of the Decline and Fall*, 1: 1026-7.

Gibbon, like most of his contemporaries, was very fond of pets.[91] A story, perhaps apocryphal, related how as a young man, hearing the cries of a dog one of the servants was beating, he intervened on behalf of the dog by beating the servant.[92] Nevertheless, he was far removed from a sentimental attitude toward animals, and in the *Decline and Fall*, while describing the sack of Rome and Italy in A.D. 410, related how the emperor Honorius, alarmed by the news that Rome was lost, regained his composure only on realizing that this was not a beloved chicken but only the capital of the world.[93] This is reminiscent of the earlier views of Richard Steele, who, despite opposing unnecessary harm to animals, which he regarded as "a great piece of cruelty, if not a kind of murder," nevertheless criticized ostentatious displays of affection toward pet animals, which he implicitly connected with possible insensitivity toward human beings.[94] A similar seeming ambiguity was evident in Gibbon's attitude toward hunting and other similar activities. He "shudder[ed] at the recollection of the bloody spectacles of the Romans; those savage combats of wild beasts, which Cicero so much despised and detested. Solitude and silence were by him preferred to these master-pieces of magnificence, horrour, and wretchedness of taste. In fact, to take delight in blood-shed is only worthy an herd of savages."[95] Nevertheless, these spectacles had their value for "naturalists."[96] Hunting was also a cruel pastime with beneficial aspects. In the Calydonian boar hunt, "Thirty or forty heroes were leagued against a hog: the brutes (not the hog) quarrelled with a lady for the head."[97] Yet Gibbon, who occasionally participated in hunting activity,[98] observed, regarding the Tartars, that hunting was a preparation for war, "and the amusements of the chace [*sic*]

91 See the stories, particularly regarding his favorite dogs, in Gibbon, *Letters*, 1: 374; 2: 65, 70, 74, 77, 371. On pet-keeping in the eighteenth century, see Tague, *Animal Companions*. On pet-keeping in the following century, see Kete, *Beast in the Boudoir*.

92 Craddock, *Young Edward Gibbon*, 132-3.

93 Gibbon, *History of the Decline and Fall*, 2: 218, note 149. For the source of this story, see Procopius, [*Works*], 2: 17.

94 Steele, *Tatler*, nos. 134 and 121 respectively.

95 Gibbon, *Essai*, 80 (English version). For the original, see the French text at ibid., 64-5: "Je rapelle en frémissant les spectacles sanglans des Romains. Le sage Ciceron les détestoit et les méprisoit. La solitude et le silence l'emportoient de beaucoup chez lui sur ces chefs-d'œuvre de magnificence, d'horreur et de mauvais gout. En effet, se plaire au carnage n'est digne que d'une troupe de sauvages."

96 Ibid., 79-82 (English), 64-6 (French). Also see Gibbon, *History of the Decline and Fall*, 1: 352-3.

97 Gibbon, *History of the Decline and Fall*, 2: 656, note 70. See also the more ambiguous criticism of hunting in ibid., 2: 185, and the praise of falconry at 2: 868. Also see Gibbon, *Letters*, 1: 352.

98 *Gibbon's Journal to January 28th, 1763*, 161-2.

serve as a prelude to the conquest of an empire."[99] Viewing hunting as both a manly and noble exercise, and consequently as preparation for war, was of course a common outlook.

Gibbon considered animals primarily as a natural resource meant for human benefit. While discussing the Arabian horses and camels (two animals also particularly noted by Raynal and Robertson), he observed that "Our toil is lessened, and our wealth is increased, by our dominion over the useful animals; and the Arabian shepherd had acquired the absolute possession of a faithful friend [the horse] and a laborious slave [the camel]."[100] While Tartar animal slaughter may have been cruel, Gibbon recognized the logistical advantage of using animals and animal products such as milk during battles, and he elaborated the important uses which animals afforded the vagrant Tartars in labor and food.[101] By moving from a seeming vegetarian position to claiming that animal suffering was necessary for human progress, Gibbon ironically undermined the theriophilic position as in essence historically unrealistic. Even more indicative was his criticism of medieval hunting laws, which enabled the animals intended for aristocratic hunters to ravage with impunity the fields industriously worked by their vassals. Regarding this he remarked how "The vague dominion, which MAN has assumed over the wild inhabitants of the earth, the air, and the waters, was confined to some fortunate individuals of the human species."[102] The moment human interests were at stake, any notion of animal worth was only as a resource or commodity. Gibbon's reference to human dominion was particularly important, and the word "vague" should be treated only as a literary flourish. Humanity's ability to cause animals with discordant natures, such as the domestic dog and cat, to live together was "proof of the empire of man over the animals."[103]

Despite Buffon's influence, purely zoological topics were of no concern for Gibbon, and he was much more interested in how the animal world could enlighten the understanding of humanity.[104] For example, he observed how the more savage a nation, the more it depended on instinct and not

99 Gibbon, *History of the Decline and Fall*, 1: 1029-30.

100 Ibid., 3: 154. See also 3: 186 and 249, for Muslim charity which descended also to animals.

101 Ibid., 1: 1026-7.

102 Ibid., 2: 481.

103 Gibbon, *English Essays*, 125 (from "Index Expurgatorius").

104 Although on at least one occasion he indulged in a short observation of the care of progeny in animals, of the type which constantly preoccupied naturalists like Buffon. See Gibbon, *History of the Decline and Fall*, 2: 808.

reason and consequently became less distinct from other nations.[105] In other words, the more human societies were akin to animals, the more their modes of life and social behavior were uniform and barbarous. The ability to change, to have a history, was essential to being truly human. What further made humanity unique was its propensity for commanding nature, and relinquishing this command jeopardized the advantages of human progress. Nature itself could never be annihilated, and even following the greatest human desolation, the potential for recommencing progress remained open to the proper civilization. Gibbon recognized this when he discussed the barbarian incursions into the Roman provinces toward the end of the fourth century. These perhaps left domesticated animals unprotected, but "the fish of the Danube would have felt more terror and distress, from the approach of a voracious pike, than from the hostile inroad of a Gothic army."[106] Progress could not be taken for granted. It was an ongoing process with no limited end in sight, requiring constant attention, in which the command of nature, and specifically animals, played a key role in providing the initial foundation of culture. Without this foundation higher civilization was impossible. Like other contemporary historians such as Herder, Raynal, and Robertson, Gibbon viewed the command of nature as vital to human progress. With all the differences between their respective intellectual concerns, all of them were influenced by Buffon's views and, to various extents, also by stadial theory. Furthermore, they all shared the predominant late eighteenth-century perception of the role the mastery of nature played in the progress of human civilization.

The common early modern attitude toward animals propounded an avoidance of *needless* cruelty to them, while nevertheless often considering such cruelty as all too necessary. In this light, Enlightenment historians, in typical eighteenth-century fashion, elaborated the Scientific Revolution's advocacy of the empirical investigation and utilization of nature. They also, however, went beyond this common outlook. Eighteenth-century conjectural history and stadial theory added a distinct historical component to the previous century's advocacy for the mastery of nature. The human command of natural resources, including animals, was increasingly viewed as a vital element of the ongoing tale of human progress. There was no comparable sustained consideration of animals in the history of historiography before the second half of the eighteenth century. Moreover, late Enlightenment

105 Ibid., 1: 1025. Gibbon in this passage continued and perceived the coupling among the Tartars of an "indolence [which] refuses to cultivate the earth," together with their military prowess.
106 Ibid., 1: 1068-9.

historical literature was increasingly preoccupied, in a somewhat pre-Marxist fashion (and obviously based on a different ethical and social outlook from Marxism), with the general material development of human societies and with the idea of material progress as the underpinning of higher cultural, social, and political progress. Conjectural history and stadial theory were essential ingredients in this increasing interest in the history of economic and material culture. It was all but inevitable that animals would become leading (passive) protagonists in this new type of historical discourse. The idea that the utilization of animals is necessary for sustaining the modern lifestyle seems to us self-evident, but this was not the case two and a half centuries ago, when the modern industrialized mass utilization of animals was only just beginning. The fact that many people today take such a view of animals for granted demonstrates both the novelty and the influence of late Enlightenment historians' treatment of this topic. It is therefore clear that late Enlightenment historical discussions of animals were one of the earliest and most significant signposts signaling the transition from the early modern to the distinctly modern consideration of animals. The old debate comparing human and animal characteristics was irrelevant to this new modern outlook. But there was still one further, even more radical, change, which Enlightenment thought implemented in its consideration of animals. This happened when the new historical outlook influenced the no less novel development of nascent economic thought. The analytical observations made by historians were, *mutatis mutandis*, transformed into the prescriptive recommendations of political economists. Yet what might be termed the economizing view of animals pervaded burgeoning modern culture in other, more diffuse, ways, not least in new pictorial depictions of animals which had begun emerging in the profane iconography of early modern art already two centuries earlier.

Part III

Art and Economics

9 Seventeenth-Century Netherlandish Paintings of Dead Animals and Changing Perceptions of Animals

The following concluding three chapters are concerned with depictions and discussions of animals in seventeenth-century art and in the history of economic thought. This might seem a rather unusual methodological pairing, but it is based on a consistent logic. The detailed explanation of this logic will become apparent in the following pages. For now, the short explanation will do, which is that both in art and in economics a new conceptualization of animals emerged, one which objectified them in a manner distinctly different from the traditional modes of discussing them which had prevailed in philosophical, literary, and scientific discourses, from antiquity till the early modern era. This change began with the increasing emphasis on the historical importance of utilizing animals, and nature in general. But in the late eighteenth century, as the modern science of political economy began to take shape, this outlook became a validation of this utilization for the sake of necessary economic progress.

The first type of animal paintings we will consider is that of dead animals. Still-life painting of dead animals emerged as an iconographic genre in early modern art, mainly in seventeenth-century Netherlandish painting. In what follows, the term "Netherlandish" will refer to both parts of that region which had effectively separated in the late sixteenth century following the war with Spain – the Southern Provinces, which remained under Catholic Spanish rule in the seventeenth century, and the new and independent Dutch Republic, with its Calvinistic religious makeup. In both countries a great artistic tradition was continued in different stylistic terms: on the one hand the Flemish baroque school under the influence of Rubens, and on the other the great Golden Age of Dutch painting. An unprecedented flourishing of profane iconographic genres developed in both countries, including genre, landscape, and still life, the latter subdivided into many sub-genres, including several types concerned with the depiction of dead animals.[1] This chapter will investigate the interaction between this artistic genre and the general history of attitudes toward animals.

1 For general introductions to seventeenth-century Netherlandish still-life painting, see Brusati, "Natural Artifice"; and Bergström, *Dutch Still-Life*. For some of the more important

Modern viewers of early modern depictions of dead animals may well feel perplexed. Such paintings often include an ostentatious display of delight in such motifs, which is quite foreign to most modern depictions of this type, for example Chaïm Soutine's expressionistic still-life paintings. Modern viewers, especially if they are sensitive to the current debate of animal rights, therefore inevitably feel a tension between the inherent beauty of such paintings and their exemplification of a seeming insensitivity to animal life and suffering. While the present discussion does not necessarily dissipate this tension, it does offer an elucidation and better understanding of it. This understanding stems from the discussion of such paintings as statements of attitudes toward animals, considered within the context of the general development of early modern attitudes toward them. This enables a better understanding of both the development of this attitude toward animals, and the nature of such paintings in themselves. Many observations regarding these paintings which might normally be considered as "subjective" attain within this context a firmer validity. It will be shown that they exemplified a surprising recognition of the value of animal life, but it will also be demonstrated that this recognition was of a very particular type.

The genre of the gamepiece, depicting animals which had been killed during hunting, was particularly popular in the early modern era.[2] One of the most interesting phenomena concerned with hunting is the aesthetic appreciation and love of nature felt by many hunters. The seeming paradox between this and the killing of animals is resolved in reality quite casually by hunters, and this is manifested by attitudes toward hunting. Throughout history hunting has been considered by many as an opportunity to come close to nature and animals, and many hunters profess feeling a close bond and love toward the animals they hunt. The love of hunters toward their victims may seem strange and even offending, but it is a very ostentatious phenomenon. Hunters rarely distinguish a conflict between love of nature and its violent exploitation. We have already noted the theory that sentimental expressions of appreciation of nature were developed mainly

among the many discussions of early modern paintings of dead animals, see Bergström, *Dutch Still-Life*, 1-41, 247-59; Craig, "Pieter Aertsen"; Freedberg, "Hidden God"; Grosjean, "Toward an Interpretation"; Koslow, *Frans Snyders*, 31-186; Moxey, "'Humanist' Market Scenes"; Schneider, *Art of the Still Life*, 7-63; and especially Sullivan, *Dutch Gamepiece*. In the years since the publication of the original version of this chapter, several other relevant publications have appeared. See in particular Cohen, "Life and Death." Also see Watson, *Back to Nature*, 208-25; and Palmeri, "Profusion." Also of interest are Cohen, "Chardin's Fur"; and Baker, "You Kill Things."

2 For a general introduction to hunting in Europe, relevant not only for medieval hunting customs, see Cummins, *Hound and the Hawk*.

by advanced urban cultures removed from direct contact with nature itself, as well as Donna Landry's theory using this claim as a justification for hunting in traditional English countryside fashion.[3] This would seem to suggest a similar tension, even though one does not have to accept it as a justification for hunting. Ernest Hemingway, one of the most famous hunters of the twentieth century, wrote that "I did not mind killing anything, any animal, if I killed it cleanly, they all had to die and my interference with the nightly and the seasonal killing that went on all the time was very minute and I had no guilty feeling at all."[4] Hunting was also common, of course, in early modern times. People of all social ranks practiced various forms of hunting, but the aristocratic types attracted the most theatrical attention. In England, for example, James I was accustomed to cutting deer's throats personally and wiping their blood on the faces of his courtiers, who were forbidden to wash it off. English ladies bathed their hands in deer's blood while these were cut open after the hunt, in the assumption that it would make them white.[5] Hunting was also popular in the seventeenth-century Netherlands, as it was in other countries.[6]

All the philosophical issues noted in the chapters above notwithstanding, the problem remains to determine to what extent, if at all, artists were aware of them, and if so, what, if any, influence they may have had on the actual painting of animal motifs. Evidence of artists directly discussing attitudes toward animals is rare. In seventeenth-century Netherlandish treatises on art, even when specific sections are devoted to animal painting, they tend to center on academic notions, such as principles of animal body depiction, or the characteristics of specific animals depicted in art, mainly horses and cattle. This is demonstrated by the chapters on animal depiction in the theoretical treatises by Karel van Mander, Chrispijn van de Passe the Younger, and Samuel van Hoogstraeten.[7] Art theoreticians, even when they were themselves painters, seem to have been more interested in the

3 See the remarks in the introduction above and also Landry, *Invention of the Countryside*.

4 Hemingway, *Green Hills*, 183. For another prominent example, see Turgenev, "Memoirs of a Sportsman," in *Memoirs*, 336-7: "To hunt with gun and dog is a proper thing in itself... But let us assume you were not born a sportsman: nevertheless, you love nature; therefore, you cannot but envy hunters like us."

5 Thomas, *Man and the Natural World*, 29.

6 On hunting in both parts of the seventeenth-century Netherlands, see the respective remarks in Balis, *Rubens' Hunting Scenes*, 22; Rosenberg, Slive, and Kuile, *Dutch Art*, 342; Sullivan, *Dutch Gamepiece*, 34-8, 87; and Zumthor, *Daily Life*, 162-3.

7 See chapter 9, "Van Beesten, Dieren, En Vogels," fols. 38-42, in "Den Gront der Edel Vry Schilder-Const," in Mander, *Het Schilder-Boeck*; Part 5, in Passe, *'t Light der Teken en Schilderkonst*; and Hoogstraeten, *Inleiding tot de Hooge Schoole*, 163-72.

depiction of the physical appearance of animals in a realistic, true-to-life manner than in the psychological characteristics of animals. This became evident in the interest of painting various subjects, including animals, according to the concept of *naer het leven* (from the life).[8] For example, in a seventeenth-century book on the lives of various artists, the Fleming Cornelis de Bie, while praising the art of the Flemish animal and still-life painter Frans Snyders, wrote: "See Snyders in the hunt [hunting scenes], how pleasantly from the life/ Shows he how to give this sweet art a great fame."[9]

At least two direct views of seventeenth-century Netherlandish artists demonstrating that they were cognizant of the contemporary debate about animals do nevertheless exist. One was the painter and entomologist Johannes Goedaert, who espoused quite extreme theriophilic views, claiming that animals had greater intelligence and morality than human beings.[10] One might also mention here the views of the prominent neo-stoic philosopher Justus Lipsius, friend of Rubens, and great lover of dogs.[11] On the other hand, Rembrandt's pupil, the painter and art theoretician Samuel van Hoogstraeten, expressed a clear predilection for the Cartesian beast-machine theory, though not in his chapter on animal painting. In discussing the difference between animals and human beings he repeated the typical Cartesian statement whereby animals "act according to their nature; more or less like a clock, which has no other impulse, than to impart the spring to the clockwork."[12] An interesting example of sensitivity to animal sentience can nonetheless be found in a late seventeenth-century Dutch emblem book titled *Het Menselyk Bedryf* (The Book of Trades), by Jan Luiken and his son Kasper Luiken. In an emblem devoted to the butcher's profession they expressed an aversion to the killing of animals (Fig. 1). Above an illustration

8 On this concept, see Swan, "Ad Vivum." Van Mander's emphasis on copying after life, and his influence in this respect on the development of seventeenth-century art, were probably less important than may seem at first sight. See Miedema, "Karel van Mander's Grondt." The term "realism" is used throughout the present chapter simply for convenience. While it has often been applied to seventeenth-century Dutch painting, recent scholarship has tended toward a more careful terminology. See Bruyn, "Turning-Point"; and Miedema, "Appreciation of Paintings," 132-3.

9 Bie, *Gulden Cabinet*, 60: "Siet SNYERS inde Jacht, hoe aerdich naer het leven/ Wist hy des' soete Const een groote faem te gheven."

10 Boas, *Happy Beast*, 50. For his work as an entomologist, see Beier, "Early Naturalists," 92.

11 Lipsius, *Discourse of Constancy*, 90-4; Papy, "Lipsius and His Dogs." On Lipsius's attitude toward animals, see also Koslow, *Frans Snyders*, 256-7; and Morford, *Stoics and Neostoics*, 4. For a different interpretation, claiming that Lipsius denied reason and passions to animals, see Harrison, "Virtues of Animals," 474.

12 Hoogstraeten, *Inleiding tot de Hooge Schoole*, 86: "zy volgen haer natuer; min noch meer als een uurwerk, dat geen andere drift heeft, dan die de veeder aen 't raederwerk meededeelt."

Figure 1 Jan and Kasper Luiken, "De Vleeshouwer", in *Het Menselyk Bedryf* (Amsterdam: Johannes and Caspaares Luiken, 1694)

of a butcher working, the motto reads: "The butcher./ Thoughtless, pays no attention," and below the illustration a short poem reads:

> The beast treads in his comrades' blood
> And knows not the fear, or to avoid,
> As like the thoughtless heart;
> All his fellow-creatures die by his side,
> And he sees the evil fruit,
> Yet does not himself take flight.[13]

13 The motto is: "De vleeshouwer./ Onbedacht, Slaat geen acht." The poem reads: "Het Beest treed in zÿn mackers bloed/ En weet van schricken noch van mÿde,/ Gelÿk het onbedacht gemoed;/ Al Sterft zÿn maasten aan zÿn zÿde,/ En dat hÿ siedt een quaade vrucht,/ Noch geeft hÿ sich niet op de vlucht." See Luiken and Luiken, *Menselyk Bedryf*, 43.

This was a rare example of direct concern for the plight of animals in a pictorial context, however, and even here the precise philosophical intent was only clear from the accompanying text. The scarcity of direct textual expressions regarding the debate on the status of animals on the part of painters is in fact unsurprising for yet another reason. The discussions of animal depiction by van Mander, van de Passe, and van Hoogstraeten, and the views of Goedaert, were expressed by painters who, with the exception of the latter, were not specialists in animal painting. Indeed, Goedaert's theriophilic sentiment may in all likelihood have arisen more from his work as an entomologist than as a painter. Nevertheless, considering the wide diffusion of the debate on animals in the seventeenth century on the one hand and the comparatively high level of education, and even at times erudition, of at least some contemporary painters on the other, it does not seem farfetched to claim that many were indeed aware of this debate. The level of literacy was indeed particularly high in seventeenth-century Holland, and we have already noted the intensive scientific effort which was put into animal experiments. One is therefore prompted to assume that if painters held any views on animals, they probably preferred expressing them in their art rather than in verbal terms. The scarcity of statements on animals on the part of painters is nevertheless surprising considering the fact that opposition to the beast-machine theory, and to cruelty to animals, was most often expressed in literary circles rather than in scientific ones, so one might have expected them also from artists. But again, it seems that most artists preferred to express their views about animals through the non-verbal medium of their paintings. In any case, claiming that paintings of animals, dead or alive, conveyed explicit philosophical content is difficult to prove, barring precise textual corroboration. A more precise attempt to understand this will be outlined in what follows. Indeed, paintings of animals might be construed as expressing either theriophilic or Cartesian meaning. For example, a painting of an animal in motion might be interpreted as depicting either a Cartesian automaton in movement or, on the contrary, a living, feeling creature. All this seems to deny any possibility of considering paintings as modes of expressions of attitudes toward animals. But this holds true only if one considers painting as a cultural medium separate from general cultural trends. This is manifestly not true. It thus becomes not only possible, but even imperative, to consider animal painting as a mode of philosophical consideration of animals, beyond the merely physical-visual aspect.

The above remarks regarding animal painting are also relevant to still-life paintings of dead animals. Animal and still-life painting were regarded in the seventeenth century as two distinct genres, and some painters developed expertise in either of the two genres, both of them, or as part of a wider repertoire.[14] According to most early modern academic theories, and well into the nineteenth century, still-life painting was regarded as probably the least prestigious of artistic genres, with landscape and animal painting ranked immediately above, then genre painting, and at the top of the iconographical hierarchy history painting, the practitioners of which enjoyed the greatest prestige and were paid the highest fees.[15] Yet the borders between genres were not always clear. Still-life paintings of dead animals often included figures of live animals as well (Fig. 2). In any event, the artistic status of a painterly genre is irrelevant to its significance as a historical source. Furthermore, it seems self-evident that paintings which center on the corpses of dead animals can and should be considered as statements, even if unconscious, of certain modes of attitudes toward animals.

Seventeenth-century Netherlandish paintings of all genres, including those of dead animals, are often interpreted in modern scholarship based on what has become known as the iconological approach. This means the attempt to determine if, and what, symbolic and moralistic messages were conveyed by the painted motifs or, in other words, whether, usually based on the abundant emblem literature of the period, certain verbal meanings could be read in the paintings. As with other genres, there are many cases in which the decision on this point is controversial.[16] Dead animal paintings with and without iconological meanings lead to a large extent to similar conclusions with respect to the attitudes toward animals they convey, though they also can differ in certain respects. This dichotomy is perceived in Dutch as well as Flemish paintings, although the iconological approach

14 In current scholarship this is quite evident, and discussions of each of these respective genres usually do not include many references to the other. This can be seen in discussions of dead-animal painting (see the relevant sources in note 1 above) on the one hand, and in the following selection of general discussions of seventeenth-century Netherlandish animal painting on the other: Balis, "Facetten van de Vlaamse Dierenschilderkunst"; *Bestiaire hollandais*, 5-7; Chong, "'In 't verbeelden van Slachtdieren'"; Heugten, "Grazing Models"; Koslow, *Frans Snyders*; Müllenmeister, *Meer und Land*, 2: 9-16; Spicer, "'De Koe voor d'aerde statt'"; Walsh, Buijsen, and Broos, *Paulus Potter*; and chapter 6, "The Historical Importance of Seventeenth-Century Dutch and Flemish Animal Painting," in Wolloch, *Subjugated Animals*, 151-76.

15 Gasten, "Dutch Still-Life." The Aristotelian gradation of natural phenomena from the vegetative, to the sensitive, and finally to the rational, mirrored in the theory of the Great Chain of Being, influenced such theories. See Koslow, *Frans Snyders*, 31-2.

16 For a good short introduction to the iconological approach by a leading iconologist, see Jongh, "Some Notes." For an introduction to this approach as pertains to still-life painting, see Jongh, "Interpretation of Still-Life." The most important critique of this approach remains Alpers, *Art of Describing*. Of related interest is also Sluijter, "Didactic and Disguised Meanings?" For a detailed discussion with further references, see Wolloch, "Iconography, Style."

Figure 2 Jan Baptist Weenix, *A Dog and a Cat near a Partially Disemboweled Deer*, 1645-1660

Rijksmuseum, Amsterdam

is in general more pertinent for the interpretation of the former. The art of the seventeenth-century Southern Provinces at its most typical was Catholic-baroque in style, full of movement and sumptuousness, while at the same time that of Holland was predominantly Protestant, less flamboyant, and more intimate, even if it shared many of the stylistic traits of Flemish painting, such as a great sensitivity to color and effects of light, including in many cases the influence of Caravaggism, and open compositions.[17] As with other secular motifs, it is likely that Flemish sixteenth-century examples of dead animal painting, brought north by refugees from Spanish persecution around 1600, served as catalysts for the development of similar paintings in Holland. The styles of such paintings in both countries developed quickly along different lines, though influences between them became gradually more and more common from about the middle of the seventeenth century, as the Golden Age of painting in both countries reached its end.[18]

Seventeenth-century dead animal painting developed to a large extent from the examples provided by sixteenth-century Flemish kitchen and market scenes, notably those by artists such as Pieter Aertsen and Joachim Beuckelaer. The painted scenes by these artists reflected the thriving condition of both the southern and the northern parts of the Netherlands, particularly before the beginning of the war with Spain. These paintings occasionally included allegorical religious criticism of abundance as *voluptas carnis* (temptation of the flesh). Such criticism was often expressed in depictions of slaughterhouses and butchers' shops. Contemporary theologians commonly regarded figures of slaughtered animals as symbols of the deaths of believers.[19] This association between animal corpses and a general concept of human death was to accompany many examples of seventeenth-century iconography of this type, whether the animal bodies were depicted as either primary or secondary compositional motifs. Depictions of slaughterhouses and butchers' shops also criticized hedonistic human behavior and therefore occasionally portrayed butchers as vulgar figures. These compositions often centered more on human death or criticism of human conduct than on animal death and suffering, though the latter meaning may also at times have been implied. In this respect, they were reminiscent of the view that harming animals was wrong, not in itself but mainly because it led to human corruption and suffering. This is a clear case regarding which at least some level of

17 Larsen, "Duality."

18 On the Flemish influence on Dutch art c. 1600, see Bruyn, "Turning-Point"; and Miedema, "Appreciation of Paintings," 128, 133.

19 Schneider, *Art of the Still Life*, 34-5, 39-41.

Figure 3 Annibale Carracci, *The Butcher's Shop*, 1680s

The Picture Gallery, Christ Church, Oxford

philosophical intent regarding the view of animals can safely be surmised, because the allegorical meaning vis-à-vis human beings was unquestionable and hinged precisely on this approach to animals. Perhaps the most famous example of a seventeenth-century painting of a butcher's shop was painted by Annibale Carracci (Fig. 3). It was relatively free of symbolic meanings, although the painter included in such scenes erotic and comic allusions, maybe even to the Commedia dell'Arte.[20] When Frans Snyders, the most important seventeenth-century Flemish still-life and animal painter, depicted a similar scene, he did not include such comic meanings. Snyders occasionally represented people eviscerating meat as a reminiscence of figures of saints' torturers, who often appeared in baroque religious compositions. He may also have intended to portray butchers as similar to anatomists. Andreas Vesalius, the famous Flemish physician and anatomist who worked in Italy, whom we have already met as an animal experimentalist, had indeed

20 Wind, "Annibale Carracci's 'Scherzo'." Another painting of a similar subject by Annibale is in the Kimbell Art Museum, Fort Worth, Texas. For Carracci, and early modern Italian dead animal painting in general, see Lodi, *Natura Morta Italiana*, 7, 26-7, 36-7, 78-81. Annibale was probably influenced by the paintings of butchers' shops by the sixteenth-century painter Bartolommeo Passarotti, which, however, were more somber in style than his own. See Freedberg, *Painting in Italy*, 572-4.

Figure 4 Frans Snyders, *Larder with a Servant*, 1635-1640

Mead Art Museum, Amherst College, Amherst, MA

recommended that anatomists study butchering techniques.[21] Again, such scenes probably symbolized human themes, yet the possibility of conscious statements regarding the animals in themselves cannot be ruled out.

All types of animals, fish, and fowl, were depicted in seventeenth-century still-life paintings. One of the most popular still-life motifs was the depiction of dead birds. In many paintings, figures of dead birds were the main motif, or they were juxtaposed with those of other dead animals. Flemish painters preferred painting large stacks of birds and other animal corpses, while Dutch painters preferred compositions on a smaller, more intimate scale. Snyders in many cases inserted baroque elements of action into his still-life paintings, such as a dog or cat sniffing food and people responding to these animals. Such elements were typical of Flemish still lifes and often had allegorical meanings connected to themes such as the confrontation between restraint and temptation or between civilization and barbarism (Fig. 4).[22] An interest

21 See Snyders's painting *A Butcher's Shop* in the Pushkin State Museum of Fine Arts, Moscow. On these points, see Koslow, *Frans Snyders*, 168, 171, 173.

22 Figures of dogs sniffing food might be connected to themes of restraint in the face of temptation or to the conflict between carnal temptation and spiritual aspiration. Snyders also invented scenes

in the animals for their own sake is difficult to prove. Snyders may have been familiar with the famous Pythagorean vegetarian principle, as described in Ovid's *Metamorphoses*, a book well-known in early modern Europe, particularly in artistic circles. Karel van Mander had included a section on this book in his treatise on painting, and in it he mentioned Pythagoras's vegetarianism and theory of metempsychosis.[23] In seventeenth-century Holland Pythagoras's name became almost a synonym for vegetarianism. This was reflected in the Dutch travel literature of the period, for example in descriptions of various exotic peoples who adhered to vegetarianism, as in the books by Jan Huygen van Linschoten, Jan Nieuhoff, and Johan (Joaness) de Laet.[24] While all this cannot prove conclusively that contemporary painters were aware of such theories or, if so, that these theories influenced their depictions of animals, alive or dead, it does make such suppositions at least likely in certain cases. Many of Snyders's, and other painters', dead-animal figures, for example, mimic human poses – lying on their backs, with heads inclined sideways, or legs crossed, at times similar to figures of fallen warriors. But again, any allusion to compassion for the animals in themselves is not certain (Figs. 2, 4).[25]

One of the climaxes of seventeenth-century dead animal painting was Rembrandt's depiction of *The Slaughtered Ox* (Fig. 5).[26] The theme of a large splayed animal carcass was not in itself original, and had appeared in Netherlandish art before.[27] But Rembrandt's painting is particularly blunt, so much so that the huge piece of meat completely dwarfs the figure of the maid in the background close by. The motif of a flayed meat carcass may

in which animals behaved riotously in the absence of human scrutiny, which might have been intended as allusions to the ideas of civilized as opposed to barbaric behavior. Such paintings became in effect animal genre paintings. Scenes of this type with dogs could symbolize various political meanings, opportunism, the consequences of neglect of duty, rebellion, etc. See ibid., 108-9, 269, 271, 273-7, 280. The preponderance of bird figures in Dutch still lifes was probably explained by the large number of migrating birds passing through the country and the frequency with which they were hunted. See Sullivan, *Dutch Gamepiece*, 33-8, 87; and Zumthor, *Daily Life*, 162.

23 For the classic statement of Pythagoras's vegetarianism and doctrine of metempsychosis, see Ovid, *Metamorphoses*, 354-66. Also see folios 119-20, in "Wtlegghingh Op den Metamorphosis," in Mander, *Het Schilder-Boeck*.

24 Linschoten, *Discourse*, 64, 71-2; Nieuhoff, *Embassy*, 57; Laet, *Empire*, 85-8.

25 Koslow, *Frans Snyders*, 42-3.

26 Rembrandt painted two versions of this painting, the one from the Louvre reproduced here, and the other in the Kelvingrove Art Gallery and Museum in Glasgow. For the purposes of the present discussion there are no significant differences between them.

27 See e.g. Joachim Beuckelaer's painting *A Slaughtered Pig* from 1563, now in the Wallraf-Richartz Museum in Cologne, which may have served as inspiration for Rembrandt's painting. See Dvořák, "Stilleben"; Moxey, *Pieter Aertsen*, 101-2; and Moxey, "'Humanist' Market Scenes," 176.

Figure 5 Rembrandt Harmenszoon van Rijn, *The Slaughtered Ox*, 1655

Musée du Louvre, Paris

have been a symbol of prudence or perhaps connected to themes of death and vanitas. The slaughtered calf was considered a symbol of the death of Christ. Rembrandt's treatment of this subject may also have been connected with the theme of the return of the prodigal son, in which the father suggests slaughtering a calf in honor of his son's return (Luke 15: 11-32). In this way, paintings of this and similar subjects can be interpreted as having a double entendre, as reminders of death on the one hand and of the need for faith and the mending of ways on the other.[28] They may also have represented the month of November, the traditional month of animal slaughter, and the festival connected with it; or else they may even have been a type of genre scene connected with this subject.[29] There were, therefore, several possibilities of embedded meanings in such compositions. Furthermore, one cannot dismiss the possibility that Rembrandt's chief interest in painting this type of subject was in fact the purely visual challenge of depicting the strong colors and dramatic composition. This would seem a more modern than early modern approach, but it cannot be ruled out and is far from improbable. There is no way to decisively decide between all these options, a point which can be made regarding most seventeenth-century paintings of dead animals. In any event, this dramatic quality is what gives these paintings added meaning, whatever interpretation one may wish to convey to them, including the possibility of sorrow at the death of the animals in itself.

Probably the most popular of seventeenth-century dead animal genres was the gamepiece. The proliferation of this type of paintings, as of all dead animal painting, may at least partially be explained by the simple fact that it was easier to paint dead animals than live ones.[30] The Flemish style of such paintings was more sumptuous than the Dutch, and it was reminiscent of mannerist kitchen and market scenes. The inclusion of live animal motifs heightened this effect. While a painter such as Snyders occasionally painted more intimate compositions with smaller numbers of animal figures, his works were predominantly Flemish and baroque in style. His pupil Jan Fyt embellished the Flemish style of gamepiece with another Flemish element, the depiction of such subjects outdoors. However, his appreciation for careful finishing of fine details also exemplified the gradual Dutch influence on

28 For these various interpretations see Bal, *Reading Rembrandt*, 382-90; Craig, "Rembrandt and 'The Slaughtered Ox'"; Sullivan, "Rembrandt's 'Self-Portrait with a Dead Bittern'"; and Wind, "Annibale Carracci's 'Scherzo'," 93.

29 For November in this context, see Zumthor, *Daily Life*, 181-2.

30 Jackson, *Great Bird Paintings*, 9.

Figure 6 Jan Fyt, *A Partridge and Small Game Birds*, 1650s

The Metropolitan Museum of Art, New York

younger Flemish artists, less dependent on the style of Rubens than the older Snyders was (Fig. 6). In any case, this Flemish style, as much if not more than the Dutch style, was to be a major influence on the painting of similar subjects in the eighteenth century.[31]

The Dutch style, at least in the first half of the seventeenth century, was usually more intimate and was characterized by a closer and more direct observation of a smaller number of objects, on many occasions in a monochrome technique, and prompted by an interest in trompe l'oeil. The illusionistic effect was heightened by omitting background scenes. The objects were no longer tilted in the direction of the spectator, as in sixteenth-century mannerist kitchen and market scenes, and the sense of intimacy was in this way heightened. Painters occasionally combined various objects, such as vegetables or larger numbers of dead animals. There were also compositions

31 Ibid., 76-7; Gerson and Kuile, *Art and Architecture*, 158-61.

Figure 7 Jan Baptist Weenix, *A Dead Partridge*, c. 1657-1660

The Mauritshuis, The Hague

which centered on the depiction of a single object, such as one dead bird (Fig. 7).[32] A different, more colorful, and compositionally complex style of Dutch gamepiece, also developed under Italian, and mainly Flemish, influence. This Flemish influence on the Dutch was stronger than the vice versa one. This style was exemplified in at least some of the gamepieces by Dutch painters such as Jan Baptist Weenix, and particularly by his son Jan Weenix, who carried this style into the eighteenth century. He often placed the dead animal figures in outside settings, another Flemish influence (Fig. 8).[33]

32 On these various points, see Bergström, *Dutch Still-Life*, 247-8; and Sullivan, *Dutch Gamepiece*, 12, 53. One of the most important precedents for a trompe l'oeil depiction of a single dead bird was the painting *Still-Life with Partridge and Iron Gloves* from 1504 (in the Alte Pinakothek, Munich), by Jacopo de' Barbari, an Italian painter who also worked in the Netherlands. See Bergström, *Dutch Still-Life*, 26-30; Jackson, *Great Bird Paintings*, 20-1; and Schneider, *Art of the Still Life*, 11-12.

33 For the Flemish influence on Dutch still lifes, see Bergström, *Dutch Still-Life*, 247-8, 252-4; and Haak, *Golden Age*, 133. On Jan Baptist Weenix, see Bergström, *Dutch Still-Life*, 252-4; Sullivan, *Dutch Gamepiece*, 4, 65 (for Goethe's appreciation of his art); and Stechow, "Jan Baptist Weenix." On Jan Weenix, see Schloss, "Early Italianate Genre."

Figure 8 Jan Weenix, *Gamepiece with a Dead Heron ("Falconer's Bag")*, 1695

The Metropolitan Museum of Art, New York

The stylistic differences between the typical Dutch and Flemish types of still-life painting were possibly also the outcome, at least to a certain extent, of the differences in the social and cultural climates in both countries. The Dutch preference for realism was probably connected with the advanced scientific development in seventeenth-century Holland. Depictions of dead animals may thus have been influenced by studies in anatomy, the development of anatomical theatres, and the importation of zoological artefacts from the Dutch overseas empire. I am not aware of any evidence that Dutch painters attended the animal experiments conducted in Leiden or elsewhere. But, as has often been observed by scholars, the scientific developments of the time, in a more general sense, may have influenced the Dutch painters' concentration on small realistic details, often at the expense of more sumptuous compositions in the style of their Flemish contemporaries. The Southern Provinces, by comparison, were a much less advanced center of scientific experimentation, and this may partially explain the limited influence there of the Dutch realistic style of painting, although the religious difference between the two countries, as already noted, was also significant.[34] The popularity of gamepieces in seventeenth-century Holland may also be partially explained by socio-economic reality. Rich burghers were prevented from participating in the more prestigious forms of aristocratic hunting. They thus may have viewed gamepiece paintings as status symbols representing the noble social status to which they aspired.[35] This in turn may have been one of the reasons for the success of Flemish-style gamepieces in Holland, as intimate paintings would have been much less effective in this ostentatious role.

In paintings in the typical intimate Dutch style, the painter could act like a sculptor and stand close, at a distance of about a meter, from the painted object, without being confronted with the major technical problem of perspective. The still-life painter could in this way decrease the spatial depth of the painting, bring the background closer to the object in the foreground, and present it as an object in itself, disregarding perspective. The viewers thus felt closer to the painted objects, as if they were within touching distance,

34 On these various points, see the remarks in Freedberg, "Science, Commerce, and Art"; George, "Alive or Dead," 183, 185; Kruyfhooft and Buys, "P. P. Rubens," 44; Parr, *Jan van Linschoten*, 196-7, 231, 282; Rupp, "Matters of Life and Death"; Scheurleer, "Early Dutch Cabinets"; Whitehead, "Georg Markgraf," 430-2; and Zumthor, *Daily Life*, 115. Alpers, *Art of Describing*, passim, presents a particular interpretation of the connection between science and art in seventeenth-century Holland. For the comparatively underdeveloped science of the seventeenth-century Southern Provinces, see Geyl, *Netherlands in the Seventeenth Century*, 2: 230.

35 Sullivan, *Dutch Gamepiece*, 38-41.

and the naturalistic illusion common in trompe l'oeil depictions was taken one step further.[36] This was bound to have a psychological effect on the viewers, whatever the intended message of the paintings. Assuming for a moment that a conscious allusion, on the parts of both artists and viewers, was made to the suffering and to the death of the animals themselves, this Dutch form of still-life representation would have tended to emphasize the more individual and "poetic" aspects of the subject. At the same time, the more flamboyant style, typical mainly of the Flemish still lifes, would have tended to emphasize the more dramatic aspects of the subject. In this way both styles complemented each other in representing different aspects of the animal tragedy.

The iconological method of interpreting seventeenth-century compositions highlights the fact that in certain Flemish, and even more so Dutch, seventeenth-century paintings, symbolic messages are clearly discerned, while in others they seem to be missing, perfunctory, or at least secondary in importance. Paintings belonging to this second group may be termed, at least in certain cases, as "decorative." In other words, they may be treated as artistic depictions which were produced and viewed mainly for their aesthetic value, with little or no deference to allegorical messages, moralistic or otherwise. Of course the aesthetic value of paintings was also important in cases where they included symbolic messages, or when they played the role of status symbols. But in cases where this was not so, this mere decorative value was even more underlined. The term "decorative" is particularly pertinent to the genre of dead animal painting. While the decorative quality of beautiful objects often depicted in Dutch still lifes, such as rare flowers or nautilus cups, is basically taken for granted, this is not the case regarding objects such as animal corpses. In this respect, early modern art differs from modern art. Modern dead animal paintings, such as those by Soutine, also, of course, have an aesthetic value. But this is not a pleasing, "decorative" value. The tendency of much of modern art to mirror the ugliness of the world, beyond anything customary in early modern art, enabled Soutine to present dead animal figures in a manner totally different from his seventeenth-century predecessors. His images are shocking, disturbing, while those of the early modern era were meant, especially when they lacked symbolic messages, to please the eye, or in other words to serve, literally, as decorations. This use of the term "decorative" may seem superficial or inadequate, but for the purposes of the present discussion it is not only convenient but also, in fact, quite accurate. The dichotomy allegorical-decorative should not,

36 Grosser, *Painter's Eye*, 16-18.

however, be taken to mean that these are necessarily always opposites, since in many seventeenth-century paintings elements of both meanings could be discerned. Whether a painting was meant to be more or less either allegorical (in the iconological sense) or decorative becomes a question of degree, and determining this relative degree is one of the intriguing challenges with which scholars of seventeenth-century Dutch art have been dealing for many years.[37]

Furthermore, the use of the term "decorative" here, from a modern perspective, seems at first sight to bring to the fore the question of the difference between the use and abuse of animals. The initial modern knee-jerk response would be a critical one – emphasizing that to depict dead animals, often after having been butchered or slaughtered, as pleasing beautiful images constitutes, by definition, insensitivity to their suffering. This argument would receive additional support from the popularity of hunting scenes in early modern art, at times painted by the same painters, for example Snyders. However, there is something un-historiographical in this type of critique. As all students of history are taught, it is precisely when the past seems to be different from the present that our scholarly senses need to become alert. To remain in such instances within the norms of modern culture therefore means failure to understand the past on its own terms. This difficulty has become a damaging trap in modern historical studies, particularly in environmental history and even more especially in the history of attitudes toward animals. The ecocritical and animal-rights perspective has much to offer, yet it is much more in place in philosophical and cultural-studies scholarship than in historical research, into which it nonetheless often spills. Consequently, in the case before us, we need to remove the question mark associated with the condemnatory *how could they* (early modern artists and spectators) have enjoyed decorative displays of dead animals, and substitute for it an exclamation mark – in other words, we need to understand that this was a prevalent aspect of early modern aesthetics and culture. As should be clear by now, this does not preclude the possibility that such paintings did indeed combine both an enjoyment of these images, together with a feeling of moral unease at the suffering of animals. Indeed, this is the whole point. The very fact that these two seemingly opposed sentiments might coexist should not in fact surprise us, and it should make us realize that there was already something quite modern

37 By the term "decorative" I do not mean to imply an interpretation such as Svetlana Alpers's anti-iconological approach, although something of her emphasis on the purely visual aspects of paintings may be relevant to such non-allegorical images.

in these paintings. The dialectic between the use and abuse of animals was an inherent and predominant aspect of these paintings, but, as we will see, it also increasingly became evident in the eighteenth-century transition to modern culture when, alongside the theriophilic sentiment which we have already described, the economic view of animals became more expressly articulated. It is easy to succumb to our contemporary view of this dialectic, specifically the frustrating realization, among modern environmentalists, that there is no way that the use of animals can be done away with completely, since it is such an essential aspect of the material underpinning of human civilization. Of course, this environmental sensitivity was not yet evident in the seventeenth or eighteenth centuries, but judging them by this standard, precisely for that reason, is historiographically futile and contributes nothing to understanding early modern attitudes toward animals in general or their artistic representation in particular. The fact is that seventeenth- and eighteenth-century audiences were often increasingly prone to at least some moral sensitivity to animal suffering but at the same time regarded artistic representations of dead animals, often mirroring ubiquitous images from daily life, as aesthetically beautiful. Our modern differentiation between the use and the abuse of animals would in this instance have seemed to them irrelevant and perhaps incomprehensible, or simply unnecessary and oversensitive.

To come back, in any event, to how seventeenth-century people themselves viewed the still-life depiction of dead animals, it is clear that not all contemporaries were pleased with these images, but for reasons which have nothing to do with proto-environmentalist sensitivities. The Dutch painter and art theoretician Gerard de Lairesse, for example, writing at the beginning of the eighteenth century, while stressing the importance of variety in painting, nevertheless opposed what he considered the abuse of painting by the depiction of unseemly objects and favored what he considered noble and edifying ones. In the realm of still life he claimed that one should avoid depicting simple things such as cabbages and herrings and should choose only such things as rare and beautiful flowers and fruits.[38] However, one should remember that de Lairesse's views exemplified an academic French influence on Dutch art in the late seventeenth century, at a time when the Golden Age of Dutch painting was in decline. Indeed, the freedom to depict almost any subject, beautiful or otherwise, in a "decorative," aesthetic manner seems to have been one of the mainstays of early modern art, particularly Dutch painting. Any preference for the depiction of beautiful

38 Lairesse, *Art of Painting*, 62-7, 421.

objects was therefore a conscious one in the seventeenth century. The point thus becomes one of determining what indeed was considered beautiful in different historical contexts, and in this respect the singularity of early modern dead animal painting is particularly highlighted. Contemporaries seem to have found depictions of dead animals aesthetically pleasing, in almost the same way that they regarded breakfast or flower pieces pleasing. It is precisely in this context that paintings of such subjects can be referred to as "decorative." From the point of view of the development of the attitude toward animals, this is particularly important.

Prima facie, this would imply that dead animal painting conveyed an inherently non-theriophilic outlook. But the situation is not so simple, and, paradoxical as this may seem, such paintings may have included a much more theriophilic sentiment, at least in a certain sense. In fact, in many cases such paintings may be regarded as anti-Cartesian statements or, in other words, as opposed to the viewpoint of the beast-machine theory. In the case of decorative dead animal paintings without overt allegorical messages, this stems from the very idea of presenting the corpse of a dead animal as a beautiful object. Of course people who enjoyed hunting, and this included most of the early modern élite, would not have been offended by this idea. The fact is, however, that in many of these paintings there was an obvious attempt to conceal or to beautify the violent aspects connected with the killing of animals.[39] This constituted a confession by way of elimination, a recognition, even if unconscious, that the lives of animals did have some value. Otherwise there would not have existed any reason to conceal violent aspects of their death, since the death of a mere machine would not have shocked a believer in the beast-machine theory. One may, of course, claim that there was no necessary connection between the way in which animal corpses were depicted and philosophical attitudes toward animals. One might regard animals as machines, and still derive enjoyment from their depiction in painting, whether alive or dead. According to such a view, any beautifying of such themes would have originated simply to avoid bad taste. However, while this may have been the case in certain instances, it seems improbable to have been the general rule. Moreover, even in paintings which did not seem to conceal the violent aspects of the animals' death, one may perceive an anti-Cartesian element. This, owing to the fact that such depictions may be claimed to have drawn their expressive and dramatic

39 On the attempt in Flemish painting to minimize the violent aspects of the animals' death, and to give the paintings a sense of delicate elegance, see Schneider, *Art of the Still Life*, 52-4. A similar claim can be made regarding many Dutch paintings.

power from the very fact that they exemplified a pre-recognition of the suffering of animals.

Dead animal paintings with ostensible allegorical meanings might also be considered as anti-Cartesian statements, albeit in a different way. Many of these paintings alluded to ideas of human death, occasionally conveying a vanitas type of moralistic warning. Often such ideas were even conveyed by explicit symbolism.[40] The figure of a dead animal might not be as direct a vanitas symbol as a human skull, but it was still clearly a reminder of death. One might therefore claim that the death of animals in such instances served only as a symbol, a means to an end, without having any value in itself. Nevertheless, it seems more probable that such depictions implied at least some meaning beyond the merely symbolic, at least some reference to the animals in themselves. Even if their death served only as a symbol, its function in this respect would have been ineffective should some inherent importance of their life and death not have been assumed. Otherwise, without some level of conceptual similarity between the symbol and what was being symbolized, this type of allegorical message would not have been cogent. The reason for this was that the life and death of an animal would have become meaningless as symbols of human life and death for anyone who regarded them as mere Cartesian automata, devoid of feeling and soul, and therefore categorically different from human beings. Did some artists or their audiences go as far as to consider such images as depictions of the tragic death of the animals in themselves? While this cannot be ruled out completely, it was in all probability a rare occurrence, if it happened at all.

Dead animal painting, however, exhibited in yet another way recognition of the value of animal life, albeit in an almost haphazard manner. This was evident in those compositions, more common in Flemish than in Dutch paintings, though appearing in both, where live animals were represented near dead ones (Figs. 2, 4). Compositions of this type in fact emphasized even more a reference to the value of animal life, whether symbolic or decorative. This, simply due to the contrast between the two states of animal existence, alive and dead. When one saw a living animal figure near a dead one, one was reminded of the fact that the latter was also alive only a short time earlier and was not originally a still object. Nevertheless, we should

40 The swan, for example, was considered a prize catch, as well as a symbol of fearlessness, and therefore the depiction of the corpse of a swan was similar to exclaiming "how the mighty have fallen." See ibid., 52; and Sullivan, "Jan Baptist Weenix," 67. The duck, especially when depicted dead, became a vanitas symbol. This stemmed from a double entendre play of words, as the Dutch word for duck is "eend," which is similar to the words "end" or "eind," meaning the English "end," occasionally in the sense of "death." See Sullivan, *Dutch Gamepiece*, 22.

remember that the value accorded to the lives of animals in dead animal paintings was in most instances small and was nowhere nearly equated with the value of human life. In cases of decorative depictions, the paintings constituted a veritable confession of the fact that the death of an animal was not something so terrible after all, because it was capable of becoming a decorative, pleasure-giving subject, if only represented in an adequately pleasing manner. And when allegorical depictions were concerned, the very fact that the death of an animal symbolized human death, and not vice versa, demonstrated that the life and death of animals were minor in importance compared to those of human beings.

Norbert Elias claimed that the medieval habit of slicing meat at the table, and occasionally serving birds without plucking their feathers, was gradually relinquished in the seventeenth century, as people began to feel less at ease with the more uncomfortable aspects connected with the production of meat. He claimed that in the development of the civilizing process, people attempted to crush any "animal" characteristics connected with themselves, including those linked with their food.[41] This type of cultural process emphasizes even more the small value related to the lives of animals, when the violent aspects of their death were neutralized and beautified in dead animal painting. This violence could be considered, in such cases, as a negative thing, not so much because of its impact on the animals as for the "shock" and discomfiture it might cause people. In this respect, the appreciation of such paintings might be construed as similar to the philosophical claim that harming animals was not a negative thing in itself but, rather, because it might lead to the corruption and harming of human beings. Therefore, while seventeenth-century Netherlandish dead animal paintings cannot be considered as purely straightforward Cartesian or anti-theriophilic statements, neither can they be interpreted, in any simple sense, as conveying the opposite, theriophilic outlook. Rather, from our perspective, they probably convey what was the popular attitude toward animals in the seventeenth century, an attitude which perceived a certain limited degree of sentience in animals, but distinctly less than in human beings. In this respect, it was an anthropocentric cultural phenomenon firmly in the mainstream of the history of attitudes toward animals. This also emphasizes the fact that the willingness of the seventeenth-century audience to accept, and even appreciate, artistic depictions of the violent death of animals, in a manner which today would most often be condemned as lacking political correctness, points to a cultural divide between that era

41 Elias, *Civilizing Process*, 95-9.

and our own, precisely of the type that should interest historians. Only by conceptually crossing this divide can we approach an understand of the meaning these paintings had in the past.

Nevertheless, it might still be claimed that at least in certain cases these paintings exhibited an even lower regard for animal suffering than the Cartesian outlook. Even when there seems to be a recognition of animal suffering, this does not necessarily imply remorse for that suffering on the part of the artists or their audience. On the contrary, assuming that such recognition existed, then the very popularity of the paintings emphasized the low regard for that suffering. Assuming that this was occasionally the case, in such instances dead animal paintings were even more insensitive to animal suffering than the Cartesian view, which at least tried to explain away this suffering, in the process implying that had this suffering existed, it would have necessarily required moral consideration. Be that as it may, whatever the type of approach to animal sentience that the artists attempted to convey, or that the viewers saw (and these were not of course always the same), seventeenth-century dead animal paintings were in accord with the common philosophical outlook which viewed nature as on the one hand something beautiful but on the other hand as something inferior to human beings and meant for their use.

While we cannot disregard the artistic value of dead animal paintings beyond the points discussed here, it seems that these points are essential for a proper understanding of these paintings and their historical meaning. This discussion began with the assumption that artistic representations of dead animals exemplified an a priori insensitivity toward animals. This assumption was at first refuted. But interestingly enough, this did not result in an interpretation of such paintings as distinct theriophilic statements, even though, in certain cases, such an intent may have been present. In general, such paintings, especially in the historical context in which they were created and viewed, seem rather to have ultimately emphasized to an even greater degree a basic anthropocentric outlook. We already emphasized at the beginning of this book that the emergence of an aesthetic sensibility to nature and animals emerged, or at least intensified, when more advanced urban populations became increasingly removed from direct contact with nature. Dead animal paintings might be viewed as an example of this cultural phenomenon, conveying the aesthetic appreciation of nature at a historical moment when even the social élite were still used to come in direct contact with animals, and with the death of animals (and indeed of human beings), much more than are modern urban populations.

Seventeenth-century Netherlandish paintings of dead animals were a clear example of the human ability to simultaneously both appreciate and exploit animals. Yet the painters of these compositions, and their audiences, seem to have been less inclined than modern artists and audiences to conceal the exploitative end of this dichotomy. This helps to explain the uneasiness which modern viewers, especially those who might be both art connoisseurs and animal lovers, might feel when confronted with such paintings. This uneasiness could be expressed by the discernment of a strange "lively" quality which pervades many of these compositions, especially the more sumptuous among them. These paintings are also impossible to ignore, due to their often compellingly high artistic value. Furthermore, they were produced in large quantities in the seventeenth century and are therefore evidence of a wide-ranging cultural approach to the viewing of animals.[42] This makes them essential to a proper understanding of the history of attitudes toward animals. It is only within this context that many of the verbal, "subjective" terms applied to these paintings receive confirmation and that a true understanding of their nature and importance becomes possible. Moreover, as we will see later on, the conceptual space which opened up between allegorical and decorative meanings in these paintings left room for one more type of consideration of animals. This was the economic approach, the subject of the following chapters. Of course, at a very basic level depicting dead animals, which were by and large meant for human consumption, was a record of an economic use of animals. Yet, as we will see, other types of animal iconography make clear that the economic aspect of consideration of animals in early modern art was a more varied and intricate phenomenon.

42 Even assuming that paintings of dead animals formed a small percentage of the overall number of paintings produced in the seventeenth century, this overall number was so large, that the number of dead animal paintings must have been large in itself. On the number of paintings painted at the time, see Miedema, "Appreciation of Paintings," 131.

10 Adam Smith and the Economic Consideration of Animals

The emphasis on animal physical and mental characteristics was central to debates about them throughout the early modern era. But toward the end of the eighteenth century a new outlook began to be developed, emphasizing the place that the use of animals had played in the historical progress of human civilization. This new type of discourse, as we have seen, was first outlined in Enlightenment historical literature. The next, and crucial, step was the transposition of this historiographical outlook on animals into the new burgeoning field of political economy. The conclusions of historical analysis were transformed into the prescriptions of economics. It was this which provided the basis for the modern consideration of animals as economic resources. As we will see in the following chapter, this changing perspective, or "economization" of animals, had in fact been gradually developing since the seventeenth century and could be noticed not just in economic discourse per se but even in artistic representations of animals. However, the most decisive stage in this process of changing views of animals was taken toward the end of the eighteenth century by the great innovator in the emergence of the modern science of economics, the Scottish philosopher Adam Smith. In the present chapter we will therefore outline his specific contribution, and in the following chapter we will look at the broader development of economic conceptions of animals. Smith's attitude toward animals has received relatively little attention from scholars.[1] Other studies, while not centering specifically on animals, have explored the relevance of his philosophy for modern environmental concerns in general.[2] These studies have recognized that Smith's direct interest in animals and environmental issues was in itself relatively minor compared to his primary philosophical concerns, but they have found certain aspects of his philosophy potentially useful for elucidating such issues from a modern perspective.

1 But see Mancilla, "Nonhuman Animals"; and Weinstein, "What My Dog Can Do." See also the short references in Thomas, *Man and the Natural World*, 28, 295; and Harwood, *Love for Animals*, 134. A different approach from the one presented here regarding the connection between debates on animals and the rise of economic theories, specifically in Germany, can be found in Bowler, "Sentient Nature."

2 See in particular Frierson, "Adam Smith"; and Frierson, "Metastandards." For a more guarded approach, see Barkdull, "How Green." Also see Callicott, "My Reply," 292-4. For environmental criticism of Smith, see e.g. Smith, "Eco-Suicidal Economics."

The intention here is not to argue with these valuable studies but, rather, to present a different outlook on Smith's attitude toward animals and nature, one grounded in a more historically contextualized approach.

In his *Theory of Moral Sentiments* Smith observed that animals had a capacity for pleasure and pain, but he also implied that they lacked consciousness of merit and demerit. Referring to *The Turkish Spy*, as we have already noted, he wrote that "Those animals ... that have been remarkably serviceable to their masters, become the objects of a very lively gratitude. We are shocked at the brutality of that officer, mentioned in the Turkish Spy, who stabbed the horse that had carried him across an arm of the sea, lest that animal should afterwards distinguish some other person by a similar adventure."[3] As we have also already seen, interest in animals was common in the Scottish Enlightenment, and Francis Hutcheson in particular, who had a significant influence on the young Smith, was specifically interested in outlining a claim for moral consideration of animals.[4] Smith himself did not evince a comparable systematic or emphatic interest in this topic. At one point he noted expressly that "As it is mankind we are chiefly connected with it must be their actions which chiefly interest our attention; Other rationall agents we are little acquainted with and the transactions which pass amongst other animals are never of so great importance to us as to attract our notice."[5] This rather commonplace statement would tend to suggest that Smith's contribution to the debate about animals was rather unexceptional. Nevertheless, his mode of discussing animals, particularly from a material-economic viewpoint, was much more original than appears initially. Smith historicized the common consideration of animals as property, emphasizing the importance of their utilization for the progress of civilization. Yet at the same time he perceived that the abuse of animals was morally objectionable.

Smith repeated many of the stock arguments in early modern discussions of animals, which essentially amounted to claiming that human beings were superior to animals, yet not in every respect. He accounted for the physical operations of bodies in Cartesian mechanical fashion, even using the famous Cartesian analogy to a watch, yet he also noted that this did not explain the operations of the mind, which were accounted for by divine wisdom.[6] Smith noted this principally about human reason, but since he

3 Smith, *Theory of Moral Sentiments*, 95.
4 Garrett, "Francis Hutcheson."
5 Smith, *Lectures on Rhetoric*, 85.
6 Smith, *Theory of Moral Sentiments*, 87.

did not deny animal rationality, albeit only a limited one, the implication was that this applied also to animals. He discussed the senses of animals, implying that he may even have conducted field observations of them. Human beings were less dependent than animals on their senses because in childhood they were in a state of dependency for a longer period of time. Through experience humans acquired enough association of ideas to enable sufficient control of the sense of sight, which meant they did not need their instincts immediately at birth.[7] Among animals there was a certain level of cooperation, for example among hounds, but immediately after they caught their prey they began fighting over it. There was no contract among animals, and their capacity for cooperation was much more limited than among humans. Human beings, on the other hand, were inherently more in need of mutual assistance. Since the propensity to barter was the basis for the development of the division of labor, this too was unique to humanity.[8] Animals often displayed greater discrepancies than humans between the talents of different individuals among the same species, but because of this lack of barter and cooperation, animals did not profit from this as a common species.[9] Also unique to humanity was the desire and ability to persuade and lead one's fellow human beings; this may have been the instinct upon which was founded the ability for speech, another important human singularity.[10]

Smith, like most of his contemporaries, regarded the difference between human beings and animals in most respects as a matter of degree, not kind. In animals as in humans, the accession of the young was reserved for the female, who put more effort into taking care of them.[11] Among animals the connection between mating couples ended when the need to take care of their offspring ended. Among human beings, however, this period of dependency of the young on their parents was the longest, resulting in very salutary effects, particularly the subjection of the passions and the yielding to parental authority, which eventually encouraged children's ability to be members of society.[12] The disposition to barter and exchange led to diversity of employment among human beings, which did not exist among animals. Animals expected to receive things from humans based on kindness, yet humans did so out of self-love and not kindness, a propensity which also

7 "Of the External Senses," in Smith, *Essays on Philosophical Subjects*, 161-8.
8 Smith, *Lectures on Jurisprudence*, 347; Smith, *Wealth of Nations*, 1: 25-6.
9 Smith, *Wealth of Nations*, 1: 29-30.
10 Smith, *Theory of Moral Sentiments*, 336.
11 Smith, *Lectures on Jurisprudence*, 27.
12 Ibid., 141-3, 438.

influenced the interactions among human beings themselves.[13] This was an allusion to the primitivistic element in much of early modern theriophily, although Smith did not share this outlook, and his view of historical progress was much more optimistic than either Montaigne's or Rousseau's.

He noted that human beings were indeed born the most delicate creatures in nature in terms of the support of life, yet among all animals "Man has received from the bounty of nature reason and ingenuity, art, contrivan<c>e, and capacity of improvement far superior to that which she has bestowed on any of the other animalls, but is at the same time in a much more helpless, and destitute condition with regard to the support and comfort of his life."[14] Human beings needed to develop clothing, cooking, and other basic technologies, but as they advanced in refinement they acquired a taste for things unnecessary for survival, such as gems. Smith continued and depicted how this process of progress led from the invention of agriculture to that of law, arts, and high culture. Clearly critical of primitivism in Rousseau's style, Smith noted that a simple laborer in an advanced civilization lived better than a king of savages. In other words, savages were superior to animals, just as civilized societies were superior to savage ones.[15] Smith connected the common early modern perception of the initial human frailty necessitating superior human invention, with specific observations on the place of animals in human material culture and economic life. Most of the aspects of his discussion of animals noted above were not particularly original. What was original was the manner in which he considered the utilization of animals as property, making it an essential part of the historical tale of human material progress.

The perception of animals as property had been common in juridical culture since antiquity, yet Smith formulated it in a novel economic-historical manner, sharing in his own way the outlook of contemporary historians like his friends Hume, Gibbon, and William Robertson. He claimed that the animals' inferiority to human beings made them the objects of human material manipulation. Killing animals was obviously less immoral than killing a human being.[16] Smith made several interesting comments related to medieval judicial proceedings against animals and inanimate objects.[17] His

13 Ibid., 348-9, 352-3; also 492-4, 570-4.

14 Ibid., 334-5.

15 Ibid., 337-41, 487-9.

16 Smith, *Theory of Moral Sentiments*, 92-3, 291.

17 The topic of animal trials has received repeated attention in modern scholarship. See e.g. Evans, *Criminal Prosecution*; Cohen, "Law, Folklore"; and Dinzelbacher, "Animal Trials." Also of interest is Mirowski, "Realms of the Natural," 456-65.

remark on ingratitude toward animals quoted above was made precisely in connection with a discussion of such proceedings.[18] Elsewhere he discussed how animals and inanimate objects which were involved in the death of a human being were punished and became deodand.[19] Such punishment "proceeds from a resentment blind and foolish indeed, but such as [illegible word in the manuscript] legislators have not neglected."[20] Nevertheless, Smith's criticism of medieval judicial conceptions of animals was not unlimited. Medieval animal trials also implied that they shared with human beings a certain judicial responsibility and were not just receptive property.

Smith repeatedly discussed animals as property, often indeed in discussions where they were not the main topic of debate and when this was simply taken for granted.[21] For example, he criticized feudal game laws which, from love of hunting and an encroaching spirit, restricted ownership of certain animals to the aristocracy; he claimed that fish and game should be common to all.[22] The injustice involved was not in the killing of the animals but in their respective distribution as property among various social groups. Similarly, the owner of a horse could seek compensation from one who had hired it in case it was damaged, but this had nothing to do with the horse itself, which was treated in this context only as property.[23] Smith repeatedly made such passing, seemingly self-evident, references to animals as property, particularly in the *Lectures on Jurisprudence* but also in the *Wealth of Nations*. For example, he discussed animals as property in connection with accession rights, as well as in many other contexts.[24] Even when Smith noted that a vegetarian diet was preferable to one with meat, this was seemingly for economic and general moral reasons, and he went no further than noting that "Decency no where requires that any man should eat butchers meat, as it in most places requires that he should wear a linen shirt or a pair of leather shoes."[25] He did write that "The trade of a butcher is a brutal and an odious business; but it is in most places more profitable than the greater part of common trades."[26]

18 Smith, *Theory of Moral Sentiments*, 94-6.

19 Smith, *Lectures on Jurisprudence*, 116-17.

20 Ibid., 139.

21 On Smith's conception of animals as property, see e.g. Witztum, "Property Rights."

22 Smith, *Lectures on Jurisprudence*, 23-4, 63. On Smith's treatment of feudal game laws and animals as property under them, see Lieberman, "Adam Smith," 229.

23 Smith, *Lectures on Jurisprudence*, 103.

24 Ibid., 27-8, and also 71-2, 82, 114, 127, 202, 208, 400. See also Smith, *Wealth of Nations*, 1: 235-55, on the utilization of cattle and other domestic animals.

25 Ibid., 2: 876; and also 1: 206-7.

26 Ibid., 1: 117-18.

This remark, however, was made in the context of a discussion about the wages of various types of labor. Forthright ethical vegetarianism was not a part of Smith's outlook.

Most significant, however, was his treatment of animals as property in relation to the four-stages theory of human progress (hunting, shepherding, agriculture, commerce). It was in this context that Smith's originality in discussing animals as property became truly evident. Stadial theory included an exposition of how human beings increased their control and utilization of nature and animals as they progressed socially and materially.[27] One of the most significant steps in this progress was the domestication of animals at the shepherding and agricultural stages. In the hunting stage the proprietary status regarding animals depended on the stage of the hunt in which they were caught.[28] This was still not a very significant utilization of animals, but that rapidly changed as human culture progressed beyond the hunting stage. By the shepherding stage the most important objects of property were animals, with property in land rising only later in the agricultural stage.[29] Smith noted in detail how this important step in human progress, so dependent on the utilization of animals, had occurred.

> But when men came to think of taming these wild animalls and bringing them up about themselves, property would necessarily be extended a great deal farther. We may consider animalls to be of three sorts. Ist, Ferae, such as are always, in a wild state. 2dly, Mansuafactae, [or those] which are those which have been tamed so as to return back to us after we have let them out of our power, and do thus habitually; tho there be others of the same sort, as stags, hares, ducks, etc. of which there are some wild and others tame. 3dly, Mansuetae, which are such as are only to be found tame, as oxen. When men first began to rear domestic animalls, they would be all under the class of the mansuefactae, as there must have been others still wild ... They considered therefore all animalls to remain in the property of him to whom they apertaind at first, as long as they retain'd the habit of returning into his power at certain times. And this continues still to be the case with regard to those animalls that are mansuetae, or what we properly call[ed] tamed.[30]

27 On how in the fourth stage of commercialism, with the rise of leisure, there were both more knowledge and more control of nature, see Berry, "Smith and Science," 134-5.

28 Smith, *Lectures on Jurisprudence*, 17-20, 459-60.

29 Lieberman, "Adam Smith on Justice," 225-6.

30 Smith, *Lectures on Jurisprudence*, 20.

Smith immediately continued to assert that this was an extension of the notion of property, which as a result of agriculture was extended to such things as land. Animals in this respect were again first and foremost treated as property. They were essential to the human need to acquire modes of subsistence, which grew in tandem with human material and social progress in general. Stadial theory emphasized the importance of modes of subsistence and consequently implied a justification for the utilization of animals irrespective of the harm done to them. As the above-quoted passage makes clear, the stadial progression also included rising control and domestication of animals, which was accompanied by an augmentation of the property rights to animals. Completely wild animals (ferae) obviously were not considered as the possession of anyone and were thus common property, belonging to the first possessor (barring social inequalities).[31] Partly domesticated animals (mansuafactae), which were let loose seasonally, assuming that they returned habitually to their purported owners, thus manifested their being the possession of the latter. Finally, completely tame and domesticated animals (mansuetae) were regarded as the absolute property of their owners.

Three types of human historical progress were thus intertwined in a mutually enhancing manner: first the stadial progress of society; second the growing control of natural resources through rising scientific and technological means, in this case the domestication of animals; and third the growing sophistication of economic and juridical culture through which human social relations progressed and adapted to growing material affluence. In other words, as society progressed and became more civilized, the use made of animals became more sophisticated and widespread and therefore also more essential and indispensable for addressing human needs. Consequently, the consideration of animals as property became more culturally entrenched from both an economic and a judicial perspective. Smith did not simply consider animals as property, as others had done before him. Property for Smith was a historicized term. It acquired different meanings in different stages of human cultural development, essentially becoming more and more significant as human society became more advanced, a point clearly evident regarding various proprietary objects, including animals. Smith's conception of animals as property thus offers a prism through which his

31 According to Smith's critique of feudalism, this is precisely what happened in the Middle Ages, when landlords unjustly dispossessed the common people and appropriated natural resources such as wild animals, which should have been common to all. See Young, "Natural Jurisprudence," 767-8; and Lieberman, "Adam Smith," 229.

general views on historical development – economic, scientific, judicial, and social – are manifested. His economic-historical analysis of this topic presented a truly original outlook on the place utilization of animals played in cultural history.

Smith did not just discuss animals in this fashion from a historical perspective, but he seemed not to perceive any ethical complications involved. Nevertheless, we here have to return to his no doubt sincere disapprobation of blatant cruelty toward animals with which we began this discussion. This was not just a single casual remark. Elsewhere he wrote expressly about "that fellow-feeling which Nature has, for the wisest purposes, implanted in man, not only towards all other men, but (though no doubt in a much weaker degree) towards all other animals. Having destined him to be the governing animal in this little world, it seems to have been her benevolent intention to inspire him with some degree of respect, even for the meanest and weakest of his subjects."[32]

Reconciling these two seemingly inconsistent aspects of Smith's attitude toward animals appears initially problematic. One can approach this as a particular manifestation of the famous Adam Smith Problem. In essence, this refers to the tension which many scholars since the nineteenth century have perceived between the emphasis on sympathy and ethics in the *Theory of Moral Sentiments*, on the one hand, and the emphasis on self-interest in the *Wealth of Nations*, on the other. Various scholars have attempted to resolve this tension and present a unified interpretation of Smith's philosophy; others have seen it as descriptive, in other words as mirroring an inherent tension in modern civilization; while still others have denied that such a problem exists, regarding it as a pseudo-problem unnecessarily invented by scholars. No definite solution has as yet been outlined to this contentious issue, and it is doubtful whether such a solution exists. The Adam Smith Problem remains essentially an interpretative issue, and in that sense exists only to the extent that scholars are willing to accept it, something that fewer and fewer among them in recent years have been willing to do in an explicit sense. In any event, attempts to reconcile Smith's outlooks in his two major works, and present them as complementary, seem to offer the most convincing approach.[33]

Nevertheless, Smith's attitude toward animals seems rather more amenable to the descriptive interpretation. In other words, it presents a tension

32 Smith, "Of the External Senses," 136.

33 A detailed bibliography of this oft-discussed topic is out of place here, but for the history of the debate see e.g. Montes, "*Das Adam Smith Problem*."

between rising modern ethical concern with animals on the one hand and increasing industrial utilization of animals on the other. This tension began emerging in Smith's time and has since exacerbated considerably. It is therefore not a tension inherent to Smith's philosophy in itself but, rather, a tension inherent to burgeoning modern industrial civilization which is evident in his writings. It seems that regarding most aspects of the Adam Smith Problem, Smith in fact considered his approaches in his two major works as offering complementary discussions which enabled a reconciliation of any seeming tensions between moral and economic concerns. Yet regarding his attitude toward animals, he probably only mirrored this tension unintentionally without offering such a reconciliation. Like all discussions of the Adam Smith Problem, this is an essentially interpretative assertion, and therefore I cannot offer distinctive proof on this point. Smith simply did not accord animals enough attention to enable a more definite interpretation. Like most of his contemporaries, he probably did not realize the potential environmental complications of industrial utilization of nature and animals. If, however, Smith's attitude toward animals basically mirrored an unresolved tension inherent to modern industrial society, this still leaves room for those claiming that his philosophical outlook in general can be viewed as philosophically consistent, if not exactly systematic. This would also suggest that the Adam Smith Problem has different thematic manifestations which are not all amenable to the same interpretative approach.

Interestingly, also, the dilemma between morality and necessity reminds us of the ethical qualms which early modern scientists felt regarding animal experimentation. In that context too, practical requirements took precedence over moral ones. The difference with economic considerations of animals, however, is that this addressed a much wider conceptualization of the use of animals, indeed one which also subsumed animal experimentation. The conjectural-history approach, which had been detailed by Buffon much more eloquently than by Smith, made clear that the utilization of animals, and nature in general, was a sine qua non of human progress, and therefore enhancing it was an ineluctable part of civilization. This also left room for at least some ethical maneuvering, since moral progress, in an expansive social and political sense, was ultimately the great aim of progress. Yet it was clear that this progress rested on a material foundation reliant on utilizing nature. The modern ecocritical preference to refer to such utilization as "domination," a word fraught with critical allusions, is therefore, from a historical perspective, irrelevant. For eighteenth-century intellectuals, no moral progress could develop in an immaterial, literally unmaterial, vacuum. While, lacking the sensitivities of later environmentalism, they did not address this point directly, their outlook inferred that no moral sensitivity toward animals could emerge before first utilizing the same animals for the progress of civilization, which would enable a sufficiently advanced culture that was capable of the idea of moral consideration of animals. A similar observation would later underline the interpretations developed by scholars such as Norbert Elias and Keith Thomas, which we noted at the beginning of this book. It is therefore evident that the historicizing and economizing view of nature and animals which emerged toward the end of the early modern era lies at the heart of the modern human-nature relationship. In a deep sense it seems insurmountable. That does not preclude the commendable aim of ameliorating environmental and animal-rights injustices. Yet it seems eminently clear that the line between unavoidable utilization and avoidable and objectionable utilization is difficult to define, but impossible to ignore. Utilizing nature and animals lies at the heart of the progress of civilization, and the term "economic" therefore encompasses the wide gamut of the Enlightenment philosophy of progress. Of course an extreme ecocritical perspective might aim for denying this idea of progress altogether. But Enlightenment philosophers would have regarded such a notion as a Rousseauist fantasy which would throw the baby out with the bathwater, denying the material basis which enabled its own existence in the first place. In other words, it would have seemed to them illogical and foolish. Whether one agrees with this or not, this is how the Enlightenment viewed the human relationship with nature and animals.

One final point we need to address is the attempts to develop an "environmentally friendly" philosophy based on Smithian concepts, and specifically regarding animals.[34] It is indeed definitely possible to utilize Smith's ethical concepts, such as the impartial spectator and sympathy, to substantiate a modern ethics of sensitivity to animal rights.[35] However, the point has to be made that Smith himself indeed did *not* do so in any sustained manner, even considering his above-quoted remark about the "much weaker degree" of "fellow-feeling" toward animals which nature had implanted in humans and which evidently also entailed a moral obligation to treat them with respect.[36] Attempts to enlist him in favor of a modern animal ethic are part of a growing recent fashion among philosophers to reinterpret or utilize the

34 See in particular Mancilla, "Nonhuman Animals." Also see Weinstein, "What My Dog Can Do"; and, on nature in general, Frierson, "Adam Smith"; and Frierson, "Metastandards."

35 For Smith's concepts of the impartial spectator, sympathy, mutual sympathy, and other aspects of his ethical system, see, for example, from a large body of literature, Griswold, *Adam Smith*; Rothschild, *Economic Sentiments*; and Raphael, *Impartial Spectator*.

36 Smith, "Of the External Senses," 136.

philosophies of various canonical figures from the history of philosophy for environmental-ethics purposes. Yet, despite its obvious merits, this remains an essentially un-historiographical endeavor and in the case of Smith is not really relevant to elucidating his own views of animals. Neither, for similar reasons, are contemporary critiques of Smith as largely responsible for the modern unrestrained and harmful use of the environment either historically accurate or helpful for a proper historical understanding.[37] What should be clear by this point of our discussion is the fact that Smith's relative disregard for animals within his moral theory was anything but fortuitous. On the contrary, it was quite typical for the Enlightenment, when growing concern with animal welfare was no doubt enhanced, yet also remained confined within clearly defined bounds which did not usually challenge what were perceived to be essential human interests. According to the common new outlook on animals toward the end of the eighteenth century, which Smith shared and in large measure expanded in economic terms, the idea that utilizing animals was necessary for human material and cultural progress was clearly perceived as such an interest. In this sense Smith was the quintessential Enlightenment philosopher.

37 Smith, "Eco-Suicidal Economics."

11 From Symbols to Commodities: The Economization of Animals in the Transition to Modernity

The trajectory of early modern conceptualizations of animals can be described, if in the broadest terms, as follows. Throughout the sixteenth, seventeenth, and early eighteenth centuries, philosophers and scientists were interested primarily in the physical and mental comparison of animal characteristics with human characteristics. The general consensus was that despite certain areas in which animals were superior to humans, the latter retained a categorical ascendancy due to their rationality and to divine directive. Theriophiles, who often also maintained a primitivistic philosophy, used those human shortcomings that did seem to exist compared with animals, not to deny human superiority, but rather to criticize and ameliorate certain human deficiencies. Among those who held different views, the Cartesian beast-machine theory of animal automatism achieved a certain popularity in the seventeenth century, but it was more notorious than famous, and modern scholars have given it disproportionate attention. By the early eighteenth century it was generally discredited. In the second half of the century, even though the traditional debate about animal characteristics periodically resurfaced, a totally new consideration emerged, which was to signal the rise of the modern debate about animals and their status vis-à-vis human beings. This consideration stemmed from the new Enlightenment discourse about conjectural history, with its emphasis on the utilization of nature and animals. It quickly played a part in the new science of political economy, with Adam Smith acting the leading role.[1] This new economic consideration of animals, however, was a much wider phenomenon, and Smith in this, as in many other respects, not only invented the terms of the new economic approach but also, as a large part of this undertaking, gave new and definite shape to concepts and ideas which had already been developing for many years before. These ideas were not confined to the realm of economic thought per se but can also be detected in other fields, and, perhaps somewhat surprisingly, already in the art of

1 For a detailed discussion of the history of economic considerations of nature, see Wolloch, *Nature in the History of Economic Thought*. See also Wolloch, "Liberal Origins"; Schabas, *Natural Origins*; and Kula, *History*.

the previous century, specifically in seventeenth-century Dutch painting. To understand this we need first to broaden our understanding of what an "economic" conception of animals can mean. Yet even a cursory view of the dead animal paintings we discussed in a previous chapter enables one avenue for doing so – it only takes a general glance at such paintings to realize that displaying animal bodies in markets and at tables is presenting them as economic commodities. However, as we will now see, there is a much wider sense in which a pictorial economization of animals occurred in the art of the period. After considering this, we will examine the development of discussions of animals in the burgeoning economic literature of the early modern era. Animals, of course, had been economic resources throughout history. Beginning with the seventeenth century, however, the distinct conception of them as such resources created the ideological background for the modern industrialized utilization of animals.

Animals in Seventeenth-Century Dutch Painting

The "animal turn" in recent years has led to a plethora of studies of the place of animals in human civilization and their relationship with human beings. Alongside studies from a philosophical, historical, and sociological perspective, art historians have also contributed new insights into this topic, the best of which have been highly cognizant of the connection between artistic representations of animals and general cultural and historical developments. Diana Donald and Martin Kemp, for example, have outlined interesting connections between artistic images and general cultural views of animals.[2] Animals, of course, have been a prime subject of art ever since prehistoric cave painting. During the Middle Ages, like many other aspects of the consideration of animals, their depiction in art was replete with religious Christian symbolism, often at the expense of attempts at empirical accuracy. Such accuracy, motivated by budding scientific interest, nevertheless became gradually evident, as was a simple enjoyment of animal imagery.[3] These tendencies were increasingly prominent in Renaissance art, not least due to the growing contact which Europeans, including painters such as Albrecht Dürer, had with exotic animals, as well as the comparable interest they

2 Donald, *Picturing Animals*; Kemp, *Human Animal*. For general surveys of animals in art, see Clark, *Animals and Men*; Dent, *Animals in Art*; and Kalof, *Looking at Animals*.

3 Resl, "Beyond the Ark." For more on medieval perceptions of animals, see Klingender, *Animals in Art*; and Salisbury, *Beast Within*.

showed in the depiction of more common animals and of the natural world in general.[4] Yet Renaissance art still often represented animals with clear symbolic import.[5]

In the seventeenth century this gradually changed. Alongside a persistence of animal symbolism, animals were increasingly depicted in a realistic and straightforward manner, concentrating on their inherent qualities, both physical and mental. They were presented as the prime iconographical themes of compositions, not just as background elements in historical painting. The beginning of this development originated in the mannerist art of the previous century, for example in Giulio Romano's equine portraits in the Sala dei Cavalli in the Palazzo del Te. These were no longer the ornate and crowded animal compositions of the early Renaissance in the style of Pisanello, nor the animal images in graphic art such as Dürer's famous *Hare*. Rather, these were full-fledged grand compositions. For that very reason they still did not convey the psychological focus which we would associate with a modern concern with animal imagery. That would need to wait for the seventeenth century.[6] This new development was represented in paintings all over Europe, but nowhere as prominently as in the Netherlands.[7] Initially it was the Antwerp School, with Rubens at its head, which led the way. Rubens himself, in his hunting scenes and other compositions, often gave animals facial and emotional intensity akin to that reserved for humans.[8] The flamboyant Baroque style of Flemish animal representation would become the main artistic focus for other contemporaries, none less so than Frans Snyders.[9] Nevertheless, if we look for a more tranquil style of seventeenth-century animal representation, one in which psychological concentration is less dramatic and more subtle and intimate, this is to be found in the Northern Provinces of the Netherlands, in the art of the Dutch Golden Age. Without discounting developments in previous eras or in other

4 For Renaissance animal paintings it is still useful to consult Howe, *Animal Life*; and Pächt, "Early Italian." For a short introduction see Dickenson, "Meticulous Depiction." Also see Lloyd, *African Animals*; and Bedini, *Pope's Elephant*. On Dürer see Koreny, *Albrecht Dürer*; Eisler, *Dürer's Animals*; and Clarke, *Rhinoceros*.

5 Janson, *Apes and Ape Lore*; Friedmann, *Symbolic Goldfinch*.

6 For the psychological innovation of seventeenth-century animal portraiture, see Senior, "Ménagerie"; and Sørensen, "Portraits."

7 Chapter 6, "The Historical Importance of Seventeenth-Century Dutch and Flemish Animal Painting," in Wolloch, *Subjugated Animals*, 151-76.

8 For Flemish animal painting in general, see Balis, "Facetten van de Vlaamse Dierenschilderkunst." For Rubens, see e.g. Kruyfhooft and Buys, "P. P. Rubens"; and Balis, *Rubens' Hunting Scenes*.

9 For Snyders, see primarily Koslow, *Frans Snyders*.

countries, the beginning of modern artistic representation of animals was nowhere more evident than there. Furthermore, situating it within the general history of attitudes toward animals, seventeenth-century Dutch animal painting has a historical significance beyond the limits of artistic representation. While historians have devoted ample attention to scientific, philosophical, and literary considerations of animals, most of them have tended to utilize contemporaneous artistic representations of animals mainly as illustrations to developments more easily discerned based on a study of traditional textual sources, whether printed or in manuscripts. Even art historians have often fallen prey to this predilection (with the exception of a few scholars such as those noted above), utilizing verbal sources mainly to help demonstrate simultaneous developments in artistic representations of animals. Yet can we also raise the possibility that certain changes in the attitude toward animals first received expression in visual, artistic works, at least in tandem with verbal modes of expression, if not *prior to* their verbal expression? I will here argue that this was indeed the case, specifically in relation to the emergence of the modern economic consideration of animals.[10] The seventeenth century was therefore simultaneously a watershed in relation to the pictorial depiction of animals, to the increasing philosophical consideration of animals, to the rise of economic thought, and specifically to the economic consideration of animals. However, as we will see, the economic discussion of animals became explicit only in the eighteenth century, while in Dutch painting this happened a century earlier, at a time when the attention devoted to animals in economic discourse was still less emphatic. We also have to constantly keep in mind that the Dutch culture which produced this painting tradition also fostered, as we have seen in the case of Christiaan Huygens, philosophical consideration of animals, as well as animal experimentation. The history of attitudes toward animals is anything but simple.

We have already discussed some basic issues related to seventeenth-century Dutch art, such as its predilection for realism and the tendency

10 When I first began studying this topic as a graduate student I very cautiously raised the possibility that seventeenth-century Netherlandish animal painting, Flemish and in particular Dutch, demonstrated a sensitivity to animals which preceded the sensitivity that historians had claimed for later philosophical discussions of animals. See Wolloch, *Animals in the Netherlands*. I did not press this point then or in my subsequent publications. However, the present chapter reintroduces this option, specifically in the context of economic considerations of animals, or at least the possibility that artistic depictions of animals kept equal pace with intellectual developments. That said, any verbal discussion of non-verbal sources always leaves room for debate and doubt.

to often convey verbal allegorical messages, which modern scholars have been studying in what has come to be known as the iconological method, concerned especially with using emblems as keys to deciphering the pictorial messages in paintings. In the 1980s and 1990s this became the topic of fierce debate. Proponents of the iconological method such as Eddy de Jongh were criticized by Svetlana Alpers, who claimed that verbal messages in Dutch painting were much less significant compared with the scientifically motivated attempt to accurately describe the minutiae of the physical world, what she termed "the art of describing." It was Italian Renaissance art, she claimed, which conveyed essentially verbal messages, not seventeenth-century Dutch art, which was meant to be seen and not read.[11] Approximately three decades later this debate seems to have settled down to a type of draw, though tending to the side of the iconologists. It is not only that museums find it easy to draw the attention of lay visitors to the compelling "stories" that Dutch paintings seem to "tell." From a more serious scholarly perspective, it is clear that verbal messages were often indeed comprehended by contemporaries. At the same time it is obvious that in many cases such messages were absent, in which case the visually "descriptive" aspect of the paintings comes to the fore, though not necessarily in the manner intended by Alpers. In the case of animal images this was of central importance, and it opened the way to seeing them as economic resources from both a symbolic and a realistic perspective. The latter was facilitated by the growing empirical view of animals, manifested not only in the accuracy of scientific illustrations but also in the growing mode among the rich of maintaining private zoos, and the importation of exotic animals by the Dutch Unified East India Company (Verenigde Oostindische Compagnie).[12] Contact with exotic animals also came through the Dutch presence in South America, as evidenced in the drawings of Maria Sibylla Merian, the zoological illustrations of Georg Markgraf, and the compositions of Albert Eckhout, which included animal themes.[13] In

11 Jongh, "Some Notes"; Jongh, "Iconological Approach"; and Jongh, "Realism and Seeming Realism." A famous example of use of the iconological method is Schama, *Embarrassment of Riches*. For Alpers's approach, see Alpers, *Art of Describing*; and Alpers, "Picturing Dutch Culture." Also of interest is Jongh, "Review of Svetlana Alpers." See also Sluijter, "Didactic and Disguised Meanings?"

12 Loisel, *Histoire des ménageries*, 2: 31-2, 49, 54-5; George, "Alive or Dead."

13 On Merian, see Davis, *Women on the Margins*, 140-202, 295-335; and Wettengl, ed., *Maria Sibylla Merian*. On Markgraf, see Whitehead, "Georg Markgraf." On both Markgraf and Eckhout, see Joppien, "Dutch Vision"; and Brienen, "From Brazil to Europe." For a well-known example of Eckhout's animal images, see his painting of two tortoises, discussed in Gelder, "Twee Braziliaanse Schildpadden."

Holland itself the empirical view of animals was testified by artists such as the bird painter Melchior d'Hondecoeter or the reptile and undergrowth painter Otto Marseus van Schrieck, both of whom bred their own animals to serve as live models.[14]

A detailed discussion of the many sub-genres of seventeenth-century Dutch animal painting is beyond our scope, and we will center only on some representative themes. Artists painted hunting scenes, landscapes with various species of animals, paintings of domestic fowl, and, as we have seen, several types of depictions of dead animals. We have also seen that substantial evidence for a direct interest of painters or art theoreticians in the contemporary philosophical debate about animals is lacking. It is not, however, in the writings of the period that we should look for evidence of sensitivity to animal sentience in paintings, but in the paintings themselves. Dutch animal painters left practically no written evidence, but their works clearly centered on portraying animals as worthy objects of attention. These included painters such as Roelant Savery, Aelbert Cuyp, Pieter van Laer, the specialist in equine motifs Philips Wouwerman, and many others.[15] The most important, in terms of his ability to convey both the physical and psychological characteristics of animals, was Paulus Potter.[16] In the nineteenth century John Ruskin would criticize Potter precisely for lack of sensitivity to the spiritual side of animal existence, but this criticism seems totally misplaced, probably coming from an anachronistic expectation for Victorian sentimentality in the style of Edwin Landseer. Modern scholarship has taken a completely different view.[17] It is true, however, that divining

14 Van Schrieck was nicknamed "Snuffelaar" (sniffer, or ferreter) because he rummaged everywhere in search of snakes, lizards, caterpillars, spiders, butterflies, and strange plants. See Houbraken, *Groote Schouburgh*, 1: 357-8; 3: 72; and Hoogstraeten, *Inleiding tot de Hooge Schoole*, 169.

15 Bialostocki, "Les bêtes et les humains"; Buysschaert, "Roelant Savery"; Wheelock, ed., *Aelbert Cuyp*; Blankert, "Over Pieter van Laer"; Duparc, "Philips Wouwerman." For Netherlandish animal painting in general, see Müllenmeister, *Meer und Land*; *Bestiaire hollandais*; and Wolloch, *Subjugated Animals*, 151-76.

16 On Potter, see first and foremost Walsh, Buijsen, and Broos, *Paulus Potter*. The main source for his life and career is the near-contemporary account in Houbraken, *Groote Schouburgh*, 2: 125-30. Among older studies, see Westrheene, *Paulus Potter*. Also see Walsh, *Paulus Potter*; Verslype, "Preliminary Study"; Turnbull, "Horse in Landscape"; and Beirne and Janssen, "Hunting Worlds."

17 Ruskin, *Modern Painters*, 5: 275-6: "Paul Potter, their [the Dutch's] best herd and cattle painter, does not care even for sheep, but only for wool; regards not cows, but cowhide... he cannot paint eyes, nor perceive any condition of an animal's mind, except its desire of grazing." This was part of a general critique of the sensitivity of Dutch painters to human psychology as well. The more accurate modern interpretation is exemplified by Ben Broos, who writes that "Potter appears to

an artist's precise attitude toward animals simply from the works of art, without written evidence, is inconclusive. In contrast with more modern animal painters such as Franz Marc, regarding early modern painters like Potter we have to rely on intuitive perception of the sensitivity to animals in his art.[18]

Potter's painting *Cat Playing with Two Dogs* (Fig. 9) is a good starting point for demonstrating his perceptive qualities in depicting animals. A cat is poised above two lapdogs, the latter two, and perhaps the former, obviously cherished pets. One dog is oblivious to the feline intruder, the other barking at it in excitement. Among modern scholars it has become common to look for some kind of symbolism in such images, perhaps political, such as divergent ways of facing political challenge, embodied by the different reactions of the dogs. In Frans Snyders's paintings, for example, riotous animals could convey the confrontation of barbarous and civilized behavior.[19] Yet there seems no reason to necessarily accept such a message here. Any idea of religious symbolism is even more remote. Both cats and dogs, however, were and remain among the most commonly represented animals in European art, and they often conveyed clear symbolic meanings.[20] In the case of the cat these were most often, though by no means always, negative. As Robert Darnton has famously shown, the resulting mistreatment of cats was still in force in the eighteenth century.[21] Both Cervantes and Rabelais, for example, conveyed such popular superstitions.[22] In Dutch art cats could be associated with the idea of sexual lust.[23] At the same time, they also on occasion conveyed more positive ideas. The most common secular allegory tied them to the sense of sight and thus also to the art of painting, though this too could have a negative import associated with pride and vanity.[24] During the French Revolution,

have had genuine compassion for his subjects, and constantly stirs the viewer's emotions with his original, almost philosophical view of farm animals." See Walsh, Buijsen, and Broos, *Paulus Potter*, 51. Also of interest are the remarks in Fromentin, *Masters of Past Time*, 117-25.

18 Marc wrote: "How does a horse see the world, how does an eagle, a deer or a dog? How impoverished, soulless is our convention of placing animals in a landscape familiar to our own eyes rather than transporting ourselves into the soul of the animal in order to divine its visual world." Quoted from Partsch, *Franz Marc*, 38. Also see Sutton, *Art, Animals, and Experience*, 1-25.

19 Koslow, *Frans Snyders*, 108-9, 269, 271, 273-7, 280.

20 For an overview, see MacDonogh, *Reigning Cats and Dogs*.

21 "Workers Revolt: The Great Cat Massacre of the Rue Saint-Séverin," in Darnton, *Great Cat Massacre*, 79-104. For a more general discussion of cultural attitudes toward cats, and their artistic representation, see Foucart-Walter and Rosenberg, *Painted Cat*.

22 Cervantes, *Adventures of Don Quixote*, 760-4; Rabelais, *Complete Works*, 640-4.

23 Jongh, "Interpretation of Still-Life."

24 Foucart-Walter and Rosenberg, *Painted Cat*, 16-17.

Figure 9 Paulus Potter, *Cat Playing with Two Dogs*, 1652

Philadelphia Museum of Art

at a time when the cruelties depicted by Darnton were still widespread, a white cat could nonetheless symbolize liberty, a motif later depicted in Gustave Courbet's *The Artist's Studio*.[25] An interesting allusion to cats was outlined by the seventeenth-century Dutch political economist Pieter de la Court, who used them as symbols of political wariness, to emphasize a commonwealth's need to opt for peaceable rather than military policies. He noted that a cat, in contrast with lions or foxes, avoided danger, provided for food and its kittens, and thus thrived more than these other animals. In the same manner nations should behave in the realm of international politics. When need be, the cat would fight more fiercely than a lion. Nevertheless, "A cat indeed is outwardly like a lion, yet she is, and will remain but a cat still; and so we who are naturally merchants, cannot be turned into soldiers." This presumption, according to de la Court, reflected what had happened to Holland under the ill-advised policy of the stadtholders, who had adopted a

25 Seibert, "Political and Pictorial Tradition."

Figure 10 Jan Steen, *Children Teaching a Cat to Dance*, also known as *The Dancing Lesson*, 1660-1679

Rijksmuseum, Amsterdam

belligerent foreign policy at the expense of the nation's inward strength.[26] In Jan Steen's jocular painting *Children Teaching a Cat to Dance* (Fig. 10), a group of children playfully though rather ruthlessly force a kitten to dance to the music of a flute accompanied by the barks of a dog. Based on the idea of the fickleness of cats, such images, depicting them in proximity to children, were a symbol, often with religious overtones, of the transience of human delights, particularly at a time of common child mortality.[27]

Dogs were much more common than cats as themes in European art. Dogs have become a metonym for the human-animal relationship in general.[28] Ancient Semitic culture, and consequently biblical references, treated the dog

26 Court, *True Interest*, 208-9, and 64, 214-15, 244-5, for other allegorical uses of the cat's figure. Also see Weststeijn, *Commercial Republicanism*, 220.

27 This was one of a group of paintings by Steen depicting children "parenting" cats. See also *The Reading Lesson* (Kunstmuseum, Basel), and *The Cat's Medicine* (San Diego Museum of Art).

28 Haraway, *Companion Species Manifesto*. For modern representations of dogs in art, see Rosenblum, *Dog in Art*.

with contempt.[29] The positive view of dogs developed later, mainly in connection with the practice of hunting by men, on the one hand, and the habit of keeping pet lapdogs by women, on the other; and in the Middle Ages and the Renaissance they became symbols of sagacity and fidelity, both conjugal and religious.[30] While, like cats, dogs were at times massacred on a large scale, this was motivated not so much by superstition but more for fear that they were conveyers of plague during epidemics.[31] By the eighteenth century the common habit of keeping pet dogs, particularly lapdogs, would raise criticism of the excessive fondness shown these pets by Englishwomen.[32] Canine lack of sexual shame played a part in the symbolic role of dogs in seventeenth-century Dutch painting, and in genre paintings promiscuous dogs served as symbolic allusions to human sexual behavior. For example, Jacob Cats, the popular Dutch author of emblem books, used a canine image as a symbol of negative desire in an emblem titled "Res immoderata cupido est" ("Desire is an immoderate thing").[33] At the same time, according to Karel van Mander dogs symbolized such things as trustworthiness, steadiness, and the sense of smell.[34] The neo-stoic philosopher Justus Lipsius loved dogs and claimed that they possessed the virtues of intelligence, physical strength, and fidelity.[35] The Dutch positive attitude toward dogs was also enhanced by the story of how a small spaniel had saved the life of William the Silent by waking him during an enemy raid.[36] Both dogs and cats could therefore have complicated symbolic meanings, the more so when they appeared together. In religious iconography, opposing a cat and a dog could occasionally symbolize the confrontation between the forces of evil (the cat) and the forces of good and loyalty to God (the dog). This was particularly common in Renaissance scenes of the Last Supper.[37]

29 Farbridge, *Studies*, 79.

30 Reuterswärd, "Dog in the Humanist's Study." Also see Sutton, *Art, Animals, and Experience*, 26-48.

31 Jenner, "Great Dog Massacre."

32 Wyett, "Lap of Luxury"; Braunschneider, "Lady and the Lapdog." Also see MacInnes, "Mastiffs and Spaniels"; Brown, "Lady, the Lapdog"; and on nineteenth-century developments, Mangum, "Dog Years." On pet keeping in general in the eighteenth century, see Tague, *Animal Companions*.

33 Cats, *Zinne-en Minne-Beelden*, 115-17. A famous example of dogs copulating as an allusion to human sexual immoderation is Frans van Mieris the Elder's painting *Brothel Scene* in the Mauritshuis.

34 "Wtbeeldinge der Figueren," in Mander, *Het Schilder-Boeck*, fol. 129.

35 Morford, *Stoics and Neostoics*, 4; Papy, "Lipsius and His Dogs."

36 For the full story, see Motley, *Rise of the Dutch Republic*, 2: 423-4.

37 Foucart-Walter and Rosenberg, *Painted Cat*, 8-12. However, the figure of the dog could also have a negative symbolic significance in the context of Last Supper scenes, as associated with Judas and as an allusion to Christ's words: "Give not that which is holy unto the dogs, neither

It is important to keep this plethora of common early modern symbolic allusions to cats and dogs in mind, because it seems doubtful that any of this is relevant to Potter's painting, despite the fact that he was probably familiar with most of these symbolic options. It is precisely this lack of symbolic content regarding such symbolically-laden animals which is so original and significant. Such symbolism would have been perfectly natural in a seventeenth-century painting, yet there is no reason here to necessarily perceive anything but a strictly *naer het leven* depiction of animal life. This approach to animal themes was emphasized further by the growing popularity of animal portraiture in the seventeenth century. While portraits of cats were rare, those of dogs became increasingly popular, and Potter himself also painted them on occasion (Fig. 11).[38] His depiction of dogs and cats therefore forms a poignant example of how animal figures which still had clear symbolic connotations could nonetheless be depicted in art without these connotations, in and for themselves. The crucial point here is that traditional symbolism could coexist with the new sensitivity to animal psychology, but the moment the symbolic element was removed, this made even more room for this novel sensitivity to animals. In the case of cats and dogs, their portrayal was that of household pets. To look at how a new perception of animals as economic resources emerged, we have to look at the depiction of other animals. The gradual decline of symbolic meanings created a conceptual space for other meanings, and while sensitivity to animal psychology was one of them, no less significant was the material-economic depiction of animals. This in itself was at times based on secular

cast ye your pearls before swine, lest they trample them under their feet, and turn again and rend you" (Matthew 7:6), an injunction long thought to refer to the unworthy reception of the Eucharist. From the sixteenth century onwards, Judas is often shown feeding a dog or cat in scenes of the Last Supper. Yet the figure of a dog in such scenes may also be a genre element functioning as a repoussoir figure. For these points, see Steinberg, "Leonardo's Last Supper," 405. A prominent example of this type of imagery is Cosimo Rosselli's fresco *The Last Supper* in the Sistine Chapel. See Lewine, *Sistine Chapel Walls*, 85-6; and Foucart-Walter and Rosenberg, *Painted Cat*, 50-1. Similar depictions include Agostino Carracci's painting in the Hermitage, St. Petersburg; and Francesco Bassano's painting in the Prado, Madrid.

38 One should note, however, that his more famous painting *The Wolf-Hound* (in the Hermitage, St. Petersburg) was both an animal portrait and a political symbol of the watchdog of Holland. See Walsh, Buijsen, and Broos, *Paulus Potter*, 141-3. For another painting, *A Dog and a Cat in a Farmyard*, previously attributed to Potter but now attributed to Dirck Wijntrack (the animal figures by Wijntrack, the landscape by Emanuel Murant), see *Old Master Paintings and Works of Art from the Bentinck-Thyssen Collection*, no. 107. The animals in this latter painting are aggressively confronting each other, yet this seems like natural canine-feline animosity, with no religious connotations.

Figure 11 Paulus Potter, *A Spaniel*, 1653

Rijksmuseum, Amsterdam

symbolism, yet there is a more general and diffuse sense in which animals were perceived in this way in seventeenth-century art.

One of the most common animals depicted in Dutch art were cattle, the staple element of the Dutch landscape and economy. More than in the landscape art of any other contemporary country, animals played a prominent role in Dutch landscape paintings. As the city burghers developed a taste for country outings, the image of idyllic nature became both a reflection of reality and a surrogate for a connection with nature which city dwellers no longer experienced in daily life.[39] Much has been written about seventeenth-century pictorial representations of cattle, including how bovine images evoked a distinct sense of economic prosperity. It is clear

39 Kettering, *Dutch Arcadia*; Vergara, *Rubens and the Poetics of Landscape*; Koslow, *Frans Snyders*, 202-3; Walsh, *Paulus Potter*, 305-6; Zumthor, *Daily Life*, 163; Freedberg, *Dutch Landscape Prints*, 61-3.

that cattle in particular were conspicuous in Dutch landscape art.[40] One painting among many serves here as an example (Fig. 12). There is a sense that this painting evokes both such ideas of economic prosperity and also sensitivity to a distinct "animal world" existence. A more direct reference to the economic role of cattle is found in Gerard ter Borch's painting *A Maid Milking a Cow in a Barn* (Fig. 13).[41] Here it is not so much a symbolic connotation as a realistic depiction of the precise manner in which cattle served human material needs. Dutch cattle were attentively groomed, cattle markets were regular events, dairy products were common, and the quality of Dutch butter was so famous that the Dutch preferred to export it and import cheaper butter from Britain for their own consumption.[42] Cattle expectedly became positive images, and for Karel van Mander the steer symbolized moderation and temperance.[43] Landscape paintings with cattle could also serve as associations of political peace, and the figure of a milking cow clearly symbolized economic prosperity.[44] Such images, nevertheless, were idealized as much as they were realistic. There were in fact serious problems with the cattle industry in seventeenth-century Holland. Pieter de la Court noted that Holland's geography and climate held many disadvantages, such as long winters, in which cattle had to be housed, thus incurring expenses and leaving less time for milking. The Dutch, he claimed, were capable of agricultural labor, but, due to the geographical and climatic qualities of the country, its economy relied principally on its fisheries, navigation, manufactures, and trade.[45] In general, the fact that the urban Dutch social elite became fond of such idealized bovine images testifies to the dialectical relationship between sentimentalization and utilization of nature, which, as we have been constantly seeing, subsequently, particularly toward the end of the eighteenth century, became a central aspect of the modern attitude toward nature. Cattle, in any case, were so ubiquitous in the Dutch landscape that it is no surprise they became artistic symbols of economic prosperity. This ubiquity suggests that this associative meaning was not necessarily in the form of intentional symbolism but was more a general common meaning which most contemporaries would have instinctively identified when contemplating landscape paintings of this type.

40 Chong, "'In 't verbeelden van Slachtdieren'"; Heugten, "Grazing Models"; Spicer, "'De Koe voor d'aerde statt'"; Gelder, "Animal and his 'Lettres de Noblesse'"; Sutton, "Noblest of Livestock."

41 On this painting, see Wheelock et al., *Gerard ter Borch*, 108-11.

42 Zumthor, *Daily Life*, 70, 72, 141, 246, 311.

43 Mander, "Wtbeeldinge der Figueren," fols. 128-9.

44 Walsh, *Paulus Potter*, 374.

45 Court, *True Interest*, 17-25, 32-3, 130-1.

Figure 12 Paulus Potter, *Cows in a Meadow Near a Farm*, 1653

Rijksmuseum, Amsterdam

An even more interesting example than cattle of this cultural and artistic phenomenon, and one hardly noticed by scholars, is the figure of pigs in seventeenth-century Dutch art. Depictions of swine could on occasion allude to the biblical story of the prodigal son (Luke 15: 11-32), who, after squandering his inheritance, was forced to eat from a trough alongside the swine. Christ's words "Give not that which is holy unto the dogs, neither cast ye your pearls before swine, lest they trample them under their feet, and turn again and rend you" (Matthew 7:6) were an injunction long thought to refer to the unworthy reception of the Eucharist. The story of the miracle of the Gadarene swine (Mark 5:1-20; Luke 8: 26-39; Matthew 8: 28-34) also reinforced negative allusions to pigs. On a more positive note, pigs could also have invoked the month of November, traditionally the annual time of animal slaughtering. Pigs were most often negative symbols of the common associations of filth, greed, sloth, and similar repulsive qualities. Yet gradually, in the seventeenth and eighteenth centuries and even more so in modern times, they acquired more positive associations, such as wealth and rural plenty, simplicity, playfulness, innocence, cleverness, and even

Figure 13 Gerard ter Borch, *A Maid Milking a Cow in a Barn*, 1652-1654

J. Paul Getty Museum, Los Angeles

good manners. They appeared as such in eighteenth-century paintings by George Morland and Thomas Gainsborough. Yet this change of approach actually began in the previous century.[46]

The motto from an emblem with an image of a pig in a popular seventeenth-century Dutch emblem book by Roemer Visscher reads: "Als my Fortuijn tot hoogheydt voert/ Soo laet ick gheen dreck ongheroert" (As my fortune soars upward/ I leave no shit unturned). The accompanying text complains of miserable people who achieve greatness and power but do not know how to govern, and pester their subjects (Fig. 14).[47] In Jan Steen's painting *A Pig Belongs in the Sty*, the use of negative swinish associations is more subtle (Fig. 15). The pig sniffing the unconscious drunkard on the left, and even more so the one peering at the raucous happenings from his sty on the right, seem the most sober figures in the composition, compared with the inebriated humans.

46 Malcolmson and Mastoris, *English Pig*; Fisher, "Politics and Porcine Representation"; Mizelle, *Pig*. On Gainsborough, see Cohen, "Thomas Gainsborough's Sensible Animals," 198-202.
47 Visscher, *Sinnepoppen*, 175.

Figure 14 Roemer Visscher, Emblem from *Sinnepoppen*, (Amsterdam: Willem Iansz., 1614)

Koninklijke Bibliotheek, The Hague

This is practically an inversion of the common anthropocentric cosmology. The humans are more animal than the animals, or perhaps the animals are more human than the humans. Or is Steen simply using common swinish associations to reinforce his habitual humoristic view of human frivolity?[48]

Most Dutch artists, however, depicted pigs primarily as domesticated animals meant to be economically utilized as food. Rembrandt's etching *The Hog* is an example, and an even more explicit example is Adriaen van Ostade's etching *The Pig Killers* (Fig. 16).[49] These types of images evoke the more quotidian aspects of the lives of pigs, who lived in close proximity to humans for millennia in a symbiosis of clear economic significance.

48 For more on this painting, see Jansen et al., *Connoisseurship*, 122-5.

49 Godefroy, *Complete Etchings*, 129-31. For Rembrandt's etching, see Schwartz, ed. *Complete Etchings*, etching B157. Also see his drawings of pigs, reproduced in Benesch and Benesch, *Drawings of Rembrandt*, 4: catalogue nos. 777-9.

Figure 15 Jan Steen, *A Pig Belongs in the Sty*, c. 1674-1678

The Mauritshuis, The Hague

Figure 16 Adriaen van Ostade, *The Pig Killers*, c. 1652

Rijksmuseum, Amsterdam

Of course the domestication of animals was one of the most crucial historical developments defining the economic relationship between them and humans, and it influenced the cultural aspect of this relationship in far-reaching ways.[50] Anthropologists such as Edmund Leach and Claude Lévi-Strauss have described how the way humans nominally address animals testifies to a taboo type of reaction – the closer they feel toward a particular animal species, the greater the need felt to retain a cosmological supremacy over this animal by addressing it in derogatory fashion. In this way the cosmological danger resulting from this proximity is disarmed. This is why the animals which in the history of domestication have proven the most benign and beneficial to human societies – dogs, pigs, and asses – have become derogatory terms.[51] The ability to represent animals such as dogs,

50 For a general survey, see Clutton-Brock, *Natural History*.

51 Leach, "Anthropological Aspects"; Lévi-Strauss, *Savage Mind*, 204-8, 213. Also see Baker, *Picturing the Beast*, 83-9. Our discussion is concerned with Western civilization, but for how

cats, or pigs in a straightforward manner, devoid of negative allusions, should therefore not be taken for granted. It serves as a cultural sign of a realistic view, both artistically and culturally, of the place such animals have in human existence, and in the case of the pig in the seventeenth century, this place was defined primarily in economic terms.

Another image, by Adriaen van Ostade's younger brother Isaac, gives the everyday aspect of this relationship a straightforward, more innocuous, depiction (Fig. 17). Whether or not this painting has an allegorical meaning is difficult to say. But even if so, one can simultaneously see it as a *naer het leven* image of the ubiquitous place pigs had in seventeenth-century life. An even more striking image is Potter's painting *Two Pigs in a Sty* (Fig. 18). If indeed "pigs belong in the sty," there was nothing demeaning or obstreperous about them. Potter seemed here to disregard any allegorical meaning, or else he used such a meaning as an excuse for his interest in the animals themselves. With his usual ability to convey animal existence, he depicted what it was like, *naer het leven*, to *be* a pig in a sty. Of course the allusion to the significance these pigs had for human economy was clear, and their probable fate loomed for the human observer, if not for the blissfully unaware animals. Yet Potter seemed to concentrate on the "piggish" moment, with no judgment or revulsion.[52]

Space permitting, we might have seen how a similar process was evident in the depiction of other animals in seventeenth-century Dutch art, and we will have more to say below about horses. It is clear, however, that Dutch artists of the period – most notably, but far from alone, Paulus Potter – consistently depicted animals in a straightforward manner, either in conjunction with traditional symbolic and allegorical modes of representation or, more significantly, without any such messages whatsoever. They did so in a realistic, *naer het leven* fashion, devoid of the more dramatic baroque elements of their Flemish contemporaries, although these too, most prominently Frans Snyders, were capable of equal insight into animal psychology. The discarding of traditional symbolism, whether religious or political, created a space for new meaning. When this was not strictly a realistic depiction of animals, often with keen psychological insight, this came either in the form of a new type of symbolic or allegorical meaning or in allusions which were direct references rather than oblique symbols – for example, Adriaen van Ostade's depiction of the slaughtering of an animal. Both of these types

this was evident in non-Western cultures, see, for example, Norton, "Chicken or the *Iegue*."

52 For this and other depictions of pigs by Potter, see Walsh, Buijsen, and Broos, *Paulus Potter*, 38, 43, 46, 105-7, 182-3.

Figure 17 Isaac van Ostade, *Peasant with a Pig*, 1644

Frans Hals Museum, Haarlem; on long-term loan from the Mauritshuis, The Hague of connotation were often economic in nature. In Potter's case animals were often depicted with piercing psychological insight. It does not seem farfetched to equate such depictions of animals with contemporaneous philosophical claims for ethical sensitivity to them. Yet in Potter's case too, the economic significance which animals had for human society was often emphatic. One could claim, of course, that depictions of farm animals in previous eras had by definition a similar economic meaning. Yet never, before seventeenth century Netherlandish painting, were images of animals given such prominent, insightful, and consistent attention in art, let alone with such talent and artistry. This novel perception of animals opened up new avenues of meaning. In this sense, at least regarding certain paintings, we can speak of the economization of animals in the art of the period. This new phenomenon in the human-animal relationship gains added significance once we remember that it was precisely at the same time that economic thought began to receive its novel modern shape, quickly according attention to the significance of natural resources, including animals. Yet this theoretical economic discussion of animals only became emphatic in the eighteenth century. Artistic expression here preceded verbal theorizing. That said, there were also some early signs in the seventeenth century that political economists were beginning, if only haphazardly, to consider animals.

In any case, even if we take a restricted view of the emergence of animal painting in the early modern era, and insist that it only consisted of a broadening of iconographical motifs, this was a dramatic enough development to warrant considering it as a new type of meaning in the depiction of animals. The comparison with political-economic discussions of animals is here particularly illuminating. Animals were of course utilized as economic resources throughout human history, and even following the terms of Enlightenment conjectural history this use may have been more significant in the earlier stages of social development. Yet this economic utilization of animals was conceptualized only toward the end of the early modern era, thousands of years after it was practically implemented. The difference between an economic practice and the conceptualization of this practice is a distinct one which is often easily overlooked, yet these are two different historical phenomena. It is this conceptualization of common practices which constituted the intellectual and cultural innovation of attitudes toward animals which was to become a central aspect of the modern attitude toward them. The emergence of animal iconography, even if we disregard for a moment the important psychological, realistic, and symbolic aspects of paintings and consider it only in and of itself, is comparable to the emergence

Figure 18 Paulus Potter, *Two Pigs in a Sty*, 1649

The Museum of Fine Arts, Houston

of economic discourses of animals, and equally significant. The moment artists and their audiences began to take for granted that depicting even such seemingly unattractive animals as pigs living on their own terms was a topic deserving of attention as the main motif of an oil painting, this in itself accorded animals a new significance and meaning. The economic perspective on animals was only one possibility in this opening up of new meanings, but it was a central one, and connected to the broader intellectual developments we have been surveying throughout these pages. In this respect, the artistic depiction of animals, including what we can term their pictorial economization, even preceded somewhat the attention to the economization of animals in intellectual discourse. Both innovations began emerging approximately at the same time, from around the late sixteenth century. Yet the artistic depiction of animals was already at its height, both quantitatively and qualitatively, by the early seventeenth century, whereas the economic discussion of animals would have to wait more than a century for Adam Smith to lift it to a similar level.

Animals in the History of Economic Thought

It is of course less problematic to define an economic consideration of animals in the context of the history of economic thought. Rather surprisingly, the development of the view of animals as economic resources has all but escaped attention in modern scholarship. This is even more surprising given the attention, most often critical, that the economic consideration has received in contemporary environmental debates. Here we will chart the growing attention to animals in nascent political economy from the seventeenth century to the nineteenth. Since this is a historical, not a philosophical, discussion, our purpose is not to condemn how animals were increasingly viewed as economic resources but to describe how this happened, and the significance this had in the shifting of the debate about animals in the transition, around the turn of the nineteenth century, between the early modern and the modern eras. Animals, of course, have been economic resources throughout history, most significantly as domesticated livestock. Yet, once again, there is a distinct difference between discussing the history of their use as such resources and the intellectual and ideological *conceptualization* of this use, a difference mirrored in the distinction between economic history and the history of economic thought. It was only in the seventeenth century that the distinct conception of animals as economic resources emerged, ultimately creating the ideological background to the modern industrialized utilization of animals, as of natural resources in general.

All this is not to say that general cultural attitudes toward animals did not emerge in an interrelated way with the very process of domestication, at stages of human social and economic development when proximity to animals was much more tangible than in more progressive stages of development. An economic relationship with animals was and is much more tangible for those living in such close proximity with them. Once again we encounter the fact that certain types of idealization of nature and animals developed precisely among the urban élites who were increasingly removed from them. It is significant that this can be said not only of aesthetic appreciation of animals but of a theoretical economic consideration of them as well.[53] Furthermore, as such direct proximity with animals became relatively less common, at least among the growing early modern urban

53 One should differentiate here between economic literature and the popular early modern genre of husbandry and agricultural literature, which was usually written by those engaged in these preoccupations.

middle class and social élite, this opened up a conceptual space for other types of changes in the view of animals, most significantly the growing perception of their economic role in human development. As the early stages of the Industrial Revolution began, precisely when Adam Smith was defining the terms of modern economic analysis, it is not surprising that the economic use of animals and the theoretical discussion of this use were changing and becoming more enhanced at approximately the same time. Economic history and the history of economic thought might be separate, yet they remain also interconnected.

Before proceeding to examine economic considerations of animals, however, one important question needs to be addressed, particularly since we are joining here a discussion of Dutch art with a discussion of the history of economic thought. That question is: why did the Dutch not develop their own tradition of economic theorizing in the early modern era, particularly in the seventeenth century, equal to that of the English? Pieter de la Court and a few others wrote about economic matters. But this was no tradition in the sense that the English were developing across the channel. Other nations might be excused, but the Dutch were the great rivals of the English for primacy in seventeenth-century trade and navigation, and indeed often outdid the English in their economic development and military exploits, even if by the end of the century English power eventually won the day. One explanation which has been offered is that it was precisely Dutch primacy in international trade which gave rise to concern in other countries, particularly England, and thus to mercantilist thought which attempted to develop means to counter this primacy.[54] In short, if developing economic thought was a response to economic problems, the Dutch of the Golden Age had no pressing need for it. But whether or not this was indeed the case, it is clear that the Dutch found other avenues for expressing economic ideas, specifically when this came to animals. Among these avenues, art held pride of place. The English, in contrast, became the leaders in the emergence of economic thought.

The term "economics" is a modern one, coined in its present sense by Alfred Marshall in the late nineteenth century. Previously, "political economy," in the most expansive cultural sense, was the preferred term. It is therefore not surprising to find, in the history of economic thought, discussions which include broad intellectual interests, animals included. The interest of economists in animals may have been desultory, but it was nonetheless highly significant. Modern scholarship has traditionally given

54 Tijn, "Dutch Economic Thought."

much attention to the place of natural resources in economic history, but not in the history of economic thought. It has only recently begun to take a consistent look at this topic, and rather expectedly has neglected to give animals adequate attention.[55] Political economists who might have been expected to show such interest disappoint on this point. The physiocrats of the eighteenth century, with their renowned emphasis on the paramount importance of agriculture, or T. R. Malthus, with his recognition of the finiteness of natural resources, are cases in point. Nevertheless, a close reading of the sources reveals a rising interest of political economists in animals, which grew in volume and sophistication, particularly from the eighteenth century.

The ancient Greek discussion of *oikonomia*, the art of managing the household, assumed that nature was capable of satisfying all human needs. The point was to manage the excess that nature made possible so as to generate surplus produce.[56] This set the tone for the anthropocentric view of nature which would predominate in later economic thought. According to Aristotle, useful modes of acquiring wealth included first and foremost the production of food, including knowledge of livestock and husbandry. This meant that maximizing the use of nature and animals was both justified and necessary. "Now if nature makes nothing incomplete, and nothing in vain, the inference must be that she has made all animals for the sake of man."[57] During the Middle Ages this outlook was to receive reinforcement from Christian cosmology. St. Thomas Aquinas, for example, repeatedly emphasized divinely ordained human supremacy over material nature, including animals. Possession of external things was natural to human beings, and it was the utilization of material nature which validated human cosmological supremacy over it.[58]

Historians of economic thought are divided on whether ancient and medieval discussions should be considered economic theory in the full sense. It is clear, however, that in seventeenth-century mercantilism we

55 I refrained from discussing animals in detail in my recent studies of nature in the history of economic thought, reserving this discussion for the present book. See Wolloch, *Nature in the History of Economic Thought*; and Wolloch, "Liberal Origins." But a serious consideration of animals has been lacking in the relevant scholarship in general. See Schabas, *Natural Origins of Economics*; Kula, *History*; Jonsson, *Enlightenment's Frontier*; Steiguer, *Origins*; Rothschild, "Maintaining (Environmental) Capital"; and, for a broader perspective, Price, "Economics, Ecology"; Mirowski, "Realms of the Natural"; and Gustafsson, "Nature and Economy."

56 Leshem, "Oikonomia Redefined." Also see Sorabji, *Animal Minds*; and Newmyer, "Animals in Ancient Philosophy."

57 Aristotle, *Politics*, 11 (quotation), 15-16.

58 Aquinas, *Political Writings*, 6, 9, 58, 68, 110, 164, 206, 251-3, 256.

first encounter a full-fledged economic theory, perhaps somewhat diffusely defined but nonetheless consistently outlined by various literati, and influential regarding public policy throughout the long seventeenth century. The empirical outlook of the Scientific Revolution was particularly influential here. The mercantilists attempted to discuss political economy with the same methodological approach of rising modern scientific praxis.[59] Furthermore, the Scientific Revolution attempted, through scientific progress, to recover humanity's prelapsarian control of nature, thus giving a new modern guise to the traditional anthropocentric classical and religious cosmology.[60] Nascent economic thought was therefore, from its inception, built on this view of justified human use of nature.

The mercantilists regarded the balance of trade as a zero-sum game, hence the assumption that the earth's resources were finite. They therefore advocated for maximal utilization of natural resources, animals included, so that their nation would maintain an edge in its contest with other nations.[61] The Spanish political economist Jerónimo de Uztáriz, for example, regarded animals mainly as economic resources, concerning himself with the proper policy regarding their importation and exportation.[62] The main center of rising economic thought, however, was Britain, due to its growing industrial, economic, and political prominence, which would continue till the early twentieth century. It was seventeenth-century English political economists who first outlined in detail the need to utilize natural resources, not least animals, even if their main concern was with timber and forestry. One of the most prominent among them, Thomas Mun, was typical in emphasizing this point, noting how England possessed both requirements for national prosperity – natural resources in themselves as well as the artificial products which resulted from the industrial development of these resources. These included "great plentie of naturall riches, both in the Sea for Fish, and on the Land for Wooll, Cattle, Corne, Lead, Tin, Iron, and many other things for Food, Rayment, & Munition; insomuch, that vpon strickt tearmes of need, this land may liue without the help of any other Nation."[63] The mercantilist who showed the greatest interest

59 Finkelstein, *Harmony and the Balance*, 101-6; Heckscher, *Mercantilism*, 2: 308-15.

60 Harrison, *Fall of Man*.

61 Finkelstein, *Harmony and the Balance*, 89-97; Jonsson, "Natural History"; Herlitz, "Art and Nature."

62 Uztáriz, *Theory and Practice*, 421-2.

63 Mun, "A Discovrse of Trade, From England vnto the East-Indies" (originally London, 1621), in McCulloch, ed., *Early English Tracts*, 1-47, at 40-1. On Mun, see Finkelstein, *Harmony and the Balance*, 74-88.

in animals, however, not just in terms of political economy, was Sir William Petty, famous primarily for developing the early form of statistical analysis known as political arithmetic. Petty composed an unpublished treatise dealing, in a surprisingly theriophilic manner, with the theory of the Great Chain of Being. He used theriophilic claims to criticize human pride, at the same time developing a rationalistic approach to religion. He claimed there was a scale of beings for terrestrial creatures and a separate one for creatures above humans. It was historical progress which had led humans to the top of the former, and therefore perhaps in the future they would likewise be able to approach the top of the higher scale.[64] Like other theriophiles, he addressed the superiority of various animals to humans in physical and sensual abilities. Yet humans remained superior in more sophisticated language or the development of religion, as well as in such things as enjoyment of sex, various technological developments, and the use of money.[65] On the latter point Petty preceded Adam Smith's later emphasis on the ability to barter, or in other words on economic activity, as a separating line between humans and animals. Petty also emphasized how, beyond these superior abilities, humans had a lot of room to improve their scientific knowledge of the world and their technological use of this knowledge.[66] Like most other theriophiles, his recognition of the qualities and sentience of animals did not lead him to relinquish an emphasis on human superiority. For Petty this was manifest specifically in furthering the technological and economic utilization of natural resources. Despite his renowned advocacy for furthering the growth of population, he recognized that the absolute number of inhabitants was less important for prosperity than the ratio between population and land and the efficiency with which natural resources were cultivated.[67]

The economic outlook on nature and animals was to be given added weight and sophistication the following century. As already noted, the physiocrats' failure to adequately discuss animals is perhaps surprising. After all, no other economic theory before or since emphasized to such an extreme extent the superiority of agriculture to all other forms of production – only agriculture, with its intimate use of natural resources, could create real surplus value, while manufactures of any different type were

64 Lewis, *William Petty*, 96-7.

65 Ibid., 108-11, 127-9, 132-3.

66 Ibid., 112-15; Lewis, "Thinking with Animals."

67 E.g., Petty, *Several Essays*, 148. Also see McCormick, *William Petty*, 179-81, 183-4; and Bonar, *Theories of Population*, 82-100.

by definition sterile.[68] Yet surprisingly, throughout their many discussions of agriculture, the physiocrats rarely addressed animals directly, and when they did do so they expectedly treated livestock animals only as economic resources related to agricultural production, disregarding other aspects of the human-animal relationship.[69] This was already an early example of the strictly material-economic approach to animals as resources which would come to dominate modern economic discourse.

This would become even more evident only slightly later in the thought of Adam Smith, as we saw in the preceding chapter. Smith was the first political economist to specifically and in detail consider the economic use of animals as a sine qua non of human social existence and progress. Subsequently, economic thought was to maintain this position almost without exceptions, even in cases of ethical sensitivity to animals. Smith's successors among the classical political economists were a case in point. David Ricardo, and as already noted also Malthus, despite a consistent advocacy for utilizing land for agricultural development, did not accord animals particular attention. In Malthus's case this is quite surprising, given his consistent interest in natural resources. This lack of attention to animals is just one reason among many to deny the current vogue for seeing Malthus as a prophet of modern environmentalism.[70] In fact, Malthus warned against the dangers of overpopulation but at the same time advocated for maximal utilization of natural resources. Like many nineteenth-century liberals he was significantly influenced by Enlightenment stadial theory, and like Smith and other proponents of stadialism he recognized that progress could only be based on mastery of natural resources, and only to the extent that these resources were amenable to cultivation.[71] While he did not spell out specifically the place of animals in such cultivation, it does not seem a far stretch to surmise that his view of them was primarily as economic resources.

Other classical political economists, however, followed Smith in according animals various levels of attention. Jean-Baptiste Say, for example, specifically noted how cockfights were barbaric and cruel and how some animals

68 For physiocracy in general, see e.g. Vardi, *Physiocrats*; and for the physiocratic view of agriculture and nature, see Meek, *Economics of Physiocracy*, 43-66, 81-4, 160, 203-30; Banzhaf, "Productive Nature"; Riskin, "'Spirit of System'"; Herlitz, "Art and Nature"; and Schaffer, "Earth's Fertility."

69 E.g., Mirabeau, *Philosophie rurale*, 1: 361-4.

70 E.g., Wells, "Resurrecting the Dismal Parson." For a more critical assessment, see Becker et al., "Malthus vs. Wordsworth." Also see Glacken, *Traces on the Rhodian Shore*, 637-54.

71 E.g., Malthus, *Essay*, 1: 40, 44, and also 60-86, 225, 385-6.

providentially managed their resources better than human spendthrifts.[72] Like Smith, however, this did not prevent him advocating for human superiority and use of animals. Humans, in contrast with animals, had future wants and therefore a level of foresight, which enabled humanity to preserve itself in the long run.[73] For Say, increased control of nature and animals meant more efficient production and hence greater profit for all members of society. "The discovery of a new mineral, animal, or vegetable, possessed of the properties of utility in a novel form, or in a greater degree of abundance or perfection, is an acquisition of the same kind [as other discoveries of natural resources]. The productive means of mankind were amplified, and a larger product rendered procurable by an equal degree of human exertion."[74]

The most sophisticated and, from our perspective, interesting case of an economic discussion of animals against a general philosophical background was provided, however, by another classical political economist – John Stuart Mill. In Mill's case the full gamut of nineteenth-century considerations of nature came into play. On the one hand was the influence of the Industrial Revolution, which included increasing forms of animal domestication and husbandry on an ever-widening scale and which ultimately was to lead to modern mass industrial utilization of animals. This, at a time when the modern affection for pets, and debates on such things as vivisection, were increasingly coming to the fore, not least in Britain, where political economy was making its most significant strides.[75] Mass utilization of animals inevitably exacerbated the traditional mistreatment of them.[76] On the other hand, romantic art and philosophy presented a more emotional contemplation of nature and animals, most often in a new modernized version of the traditional theriophilic approach, which now was less concerned with the classical use of animals as exempla for human conduct and more centered on a direct concern with moral obligations toward animals.[77] Moreover, in the second half of the century, when Mill was repeatedly revising his *Principles of Political Economy*, the most comprehensive book on the subject since Smith's *Wealth of Nations*, the ideas of Darwin would

72 Say, *Treatise*, 403, 459.

73 Ibid., 372, 376.

74 Ibid., 299.

75 Ritvo, *Animal Estate*; Turner, *Reckoning with the Beast*; Kean, *Animal Rights*; Kete, ed., *Cultural History*; Kete, *Beast in the Boudoir*.

76 Griffin, "Animal Maiming."

77 Perkins, *Romanticism and Animal Rights*; Kenyon-Jones, *Kindred Brutes*. For developments in America, see Davis, *Gospel of Kindness*.

shake the foundations of the traditional view of human singularity, even if the implications this would have for the ethical consideration of animals were, and remain, debatable.[78] All of these influences might explain Mill's seemingly inconsistent view of animals.

Mill, of course, as a utilitarian philosopher following in the footsteps of his father and of Jeremy Bentham, approvingly quoted the latter's famous claim that "The question is not, Can they [animals] reason? nor, Can they Talk? but, Can they suffer?"[79] He criticized William Whewell's objection to Bentham's claim and to the idea that humans owed animals moral consideration.[80] The utilitarian principle of the least pain and the greatest happiness for the greatest number was to be extended, according to Mill, as far as possible to all sentient creatures, not just to human beings, a point which later utilitarians, most famously Peter Singer, would extend to very controversial extremes.[81] Most tellingly, it was precisely in *Principles of Political Economy* that Mill squarely addressed the social duty to provide animals with ethically sensitive treatment. Among the duties of government which superseded the principle of laissez faire was protection of the weak, and among these Mill included animals. Punishment of crimes against animals in Britain was in his opinion insufficient. Laws on this point were concerned too much with the adverse effect that hurting animals had on human interests and not enough with the intrinsic damage done to the animals themselves.[82] This was a significant point. Many philosophers, before and after, have taken the view that according animals moral attention was important first and foremost because it guaranteed the proper treatment of human beings, not because of the moral worth of animals in themselves. Immanuel Kant was a prominent example among many, noting specifically that duties toward animals derived above all from duties toward human beings.[83] Mill, however, departed from this common outlook, preferring to see animals as ends in themselves. This was part of his broad liberal philosophy, encompassing sensitivity to the plight of women, slaves, and other human disadvantaged and including in this purview animals. This

78 For different perspectives, compare Rachels, *Created from Animals*; and Preece, "Darwinism." For an environmentally oriented discussion of nineteenth-century views of nature, including the influence of Darwin, see also LaFreniere, *Decline of Nature*, 223-59.

79 Bentham, *Introduction*, 282-3.

80 "Whewell on Moral Philosophy," in Mill, *Essays on Ethics*, 185-7.

81 "Utilitarianism," in Mill, *Essays on Ethics*, 214. See also Singer, *Animal Liberation*.

82 Mill, *Principles of Political Economy*, 2: 952.

83 Kant, *Lectures on Ethics*, 239-41.

type of consideration of animals seems to have been a family tradition, shared by his daughter-in-law Helen Taylor.[84]

Nevertheless, Mill was a classical political economist, and as such he accepted the basic assumption that utilization of natural resources was vital to the process of production and to human material and social progress. Both in the essay *Nature* and repeatedly throughout the lengthy text of *Principles of Political Economy*, he voiced the unambiguous admonition to utilize natural resources, primarily, though not exclusively, in agriculture. The discussion in the *Principles* was indeed framed by this assertion, beginning and ending with exhortations to maximize the human use of nature.[85] Whether purposefully or not, Mill did not specifically address the use of animals in this context, though he was probably aware that this posed a contrast to his admonition to treat them in a morally sensitive manner. From a broader perspective, however, he was well aware of the contradiction, so endemic to the culture of the nineteenth century, between the economic values of industrialization and the aesthetic contemplation of nature advocated by Romanticism. It is of course broadly acknowledged that Mill struggled to find the proper balance between the values of capitalism and socialism. In this as in other respects, his intellectual outlook was a prescient outline of problems and dilemmas which today still beset modern civilization. The same can also be said of his view of nature. Precisely in the text of *Principles of Political Economy*, amidst constant references to the benefits of utilizing nature, Mill suddenly, in a remarkable passage, noted the need to curtail human development before all of nature was utilized. Only then would some portions of nature, including animal nature, be left for the humanly necessary need for solitary contemplation.

> Nor is there much satisfaction in contemplating the world with nothing left to the spontaneous activity of nature; with every rood of land brought into cultivation, which is capable of growing food for human beings; every flowery waste or natural pasture ploughed up, all quadrupeds or birds which are not domesticated for man's use exterminated as his rivals for food, every hedgerow or superfluous tree rooted out, and scarcely a place left where a wild shrub or flower could grow without being eradicated as a weed in the name of improved agriculture.[86]

84 Taylor, "Few Words."

85 Mill, *Principles of Political Economy*, 1: 3; 2: 971; "Nature," in Mill, *Essays on Ethics*, 373-402. For Mill's general consideration of nature, see Schabas, *Natural Origins of Economics*, 125-33.

86 Mill, *Principles of Political Economy*, 2: 756. On this passage see also Evensky, *Adam Smith's Moral Philosophy*, 308-12; Steiguer, *Origins*, 149-62; Parham, "What is (Ecological) 'Nature'?"; and O'Connor, "John Stuart Mill's Utilitarianism."

This was a crucial point. Human beings should decline to push the utilization of nature and animals to its maximal limits. Yet this was also only one passage in a lengthy text many hundreds of pages long, which otherwise advocated uninhibited use of natural resources. Mill offered no real solution to this quandary, though this did point to a realization on his part of a vital modern dialectical tension between economic development and the cultural, spiritual, and moral price which human beings paid for this development, as necessary as it was. Once again we encounter here, perhaps in its most emphatic form, a recurrent leitmotif of our discussion: how the modern sentimentalization and aesthetization of nature was developed precisely by urban populations removed from direct contact with it.

Mill's views were in many respects a continuation, *mutatis mutandis*, of the views of his Enlightenment predecessors, and regarding economics specifically of Adam Smith. Since we are concerned here with discussing how the Enlightenment view of animals in the long eighteenth century transformed earlier notions into what would become the new modern view of animals, it makes sense to follow the thread of this development into the nineteenth century. Mill's intellectual candor in confronting the intransigent problem of the human-nature relationship was rarely seconded by contemporaneous or subsequent economists, most of whom preferred more straightforward admonitions for the use of nature and animals. While a detailed overview is beyond the scope of our discussion, it should be noted that, with almost no exceptions, modern economic thought has tended to emphasize the need to utilize nature and animals rather than to conserve them or to accord them ethical preference over economic use. Taking a look at a few other nineteenth-century and even later economic views of animals seems in order, even if it pushes the temporal boundaries of our discussion, because a central point we are making here is that the changing conceptualization of animals which occurred, primarily in historical and then in economic discourses, in the late eighteenth and early nineteenth centuries, was the turning point in the transition from the early modern to the modern consideration of animals. Several key examples from the more modern economic literature can therefore help substantiate the modernity of this transition.

Even Marxist thought did not offer an alternative to the economic view of animals. In contrast with modern leftist and neo-Marxist environmental thinking, Karl Marx himself preferred human socio-economic needs to any kind of environmentalism in the modern sense, as did Friedrich Engels. Animals manifested the human labor which had been invested in them and were first and foremost considered as economic factors, and any injustice in their use was not toward the animals in themselves but, rather, in the social distribution of the material goods which resulted from their utilization.[87] On this point Marxist thought was not much different from liberal economics. More moderate nineteenth-century socialists also, unsurprisingly, shared a similar outlook. Simonde de Sismondi indicatively claimed that a man stranded on an island filled with animals and fertile land would remain poor unless he applied industry to using these resources.

> But, if his industry enabled him to catch some of the animals that wander in his woods; and if, instead of consuming them immediately, he reserves them for his future wants; if, in this interval, he gets them tamed and multiplied, so that he can live on their milk, or associate them to his labour, he is then beginning to acquire wealth, because labour has gained him the possession of these animals, and a fresh labour has rendered them domestic.[88]

The perspective of the twenty-first century might make us expect a leftist outlook to include an environmental and animal-rights perspective within the general purview of a social-justice agenda, but this was not in fact the case in the nineteenth century. Another even more poignant example was Engels, who accorded animals more attention than either Marx or Sismondi and who anticipated the insights of modern environmental history when he noted how goats had prevented the regeneration of forests in Greece, or how in St. Helena the goats and pigs brought by navigators almost exterminated the local vegetation, making room for plants subsequently introduced by colonists. Animals, however, did such things unintentionally, while human beings did so deliberately. Human action transfers "useful plants and domestic animals from one country to another and thus changes the flora and fauna of whole continents. More than this. Through artificial breeding, both plants and animals are so changed by the hand of man that they become unrecognizable."[89] Yet Engels was no environmentalist in the modern sense, and he ultimately advocated for maximizing the human utilization of nature, paying due note to proper social distribution of the resultant wealth. When social production and distribution would be sufficiently realized, humanity

87 Parsons, ed., *Marx and Engels*, 132-3, 151, 153; Engels, *Dialectics of Nature*, 47-9, 231-2, 239-42. See also Benton, "Engels."

88 Simonde de Sismondi, *New Principles*, 61.

89 Engels, *Dialectics of Nature*, 239-40, and also 241-2, where the unforeseen consequences of deforestation are also discussed. Such processes have been a major concern for modern environmental historians. See e.g. Crosby, *Ecological Imperialism*.

would emerge beyond the world not just biologically but also socially.[90] Repeatedly mentioning Darwin, though in a sense more reminiscent of Lamarck's theory of acquired characteristics, Engels outlined a conjectural history of how human beings' use of their hands enabled the development of other organs, and eventually the social implications of enhanced physical abilities. "The mastery over nature, which begins with the development of the hand, with labour, widened man's horizon at every new advance."[91] This left very little room for moral sensitivity to animals. Engels's instrumental discussion of animals in the context of a conjectural-history discussion of human development was itself one proof among many of the varied influence of the Enlightenment in general, and classical political economy in particular, on socialist and Marxist thought.

Coming back to the advocates of liberal economics, the classical political economists were succeeded by the neo-classicists, the progenitors of the marginal revolution in economic thought, with its increasing emphasis on marginal utility, supply and demand, and an ever-increasing stress on maximizing utility much beyond Smith's earlier and more ethically nuanced outlook. Unsurprisingly, this economic approach made the instrumental view of animals absolutely explicit, as evidenced by Léon Walras, one of the leading neo-classicists, who distinguished between persons and things, the latter including inanimate objects and animals.

> Whatsoever is not conscious of itself and not master of itself is a thing. Whatsoever is conscious of itself and master of itself is a person. Man, being both self-conscious and self-directing is a person. Man alone is a person, minerals, plants and animals are things. From the rational point of view, the purpose of things is under the dominion of the purpose of persons. Since a thing is neither self-conscious nor self-directing, it is not responsible for the pursuit of its ends or for the fulfilment of its destiny. Being just as incapable of evil as it is of good, it is always completely innocent and can be likened to a pure mechanism. In this respect animals do not differ from minerals and plants: an animal's instinct is a blind and ineluctable force like any other natural force. A person, on the other hand, just because he is conscious of himself and master of himself, is charged with responsibility for the pursuit of his ends and for the fulfilment of his destiny. If he succeeds, he has merit; if not, he takes blame. He has, therefore, an unlimited faculty for subordinating the purpose of things

90 Engels, *Dialectics of Nature*, 47-9.

91 Ibid., 231-2.

> to his own purpose. This faculty, in all its length and breadth, is invested with a particular character. It is not only a moral power, it is a right. This is the basis of the right of persons over things.[92]

Gone, in one fell swoop, are centuries of careful philosophizing on the relative similarities and differences between animals and human beings. Walras continued to note that persons, in contrast with their relationship with animals, did not hold rights over each other. The totality of phenomena existing between people he termed "institutions," whereas those between people and things were "industry." "The theory of industry is called *applied science* or *art*; the theory of institutions *moral science* or *ethics*."[93] Needless to say, this provided the ethical justification for unbridled maximal economic utilization of natural resources, with animals consequently considered precisely akin to inanimate resources, no more and no less. It was this type of discussion which was later to lead Aldo Leopold, one of the founding fathers of modern environmentalism, to become disillusioned with the ability of economists to deal with a true appreciation of nature and provide a way to conserve it.[94] Mill had at least confronted the problem, even if few other economists have followed suit. One does not need to hold distinct environmental views in Leopold's style to recognize the extreme nature of Walras's view of animals. But perhaps the true potency of this view was that it was not really extreme, perhaps only in its style rather than its content. Walras was simply stating the obvious in terms of economic considerations of animals. What is interesting is how he utilized what seems at first sight the common type of traditional arguments, going back to the early modern era and even to classical philosophy, about the mental differences between animals and human beings as a basis for an anthropocentric argument about the place of the former in the civilization of the latter. But a close reading of this passage reveals the stark difference between it and seemingly similar preceding types of discussions of animals. There is nothing here reminiscent of early modern theriophilic comparisons, nor even of the Cartesian beast-machine, which at least tacitly acknowledged that had animals possessed sentience they would have been entitled to moral consideration. Walras was in all probability well aware of Darwin's theory of evolution when he wrote this passage, but if so this did not lead him to any claim of proximity or similarity between humans and animals. If anything, this passage was a

92 Walras, *Elements of Pure Economics*, 62-3.

93 Ibid.

94 Goodwin, "Ecologist Meets Economics."

modernized version of the theory of the Great Chain of Being, but, in contrast with the earlier versions of this theory, it emphasized the disruption more than the continuum in this chain and consequently accentuated the divide between human beings and the rest of nature in a manner unprecedented in similar early modern discussions. The truly novel aspect of Walras's claim was its stark rationalistic economic argumentation, one which would quickly come to typify the attitude toward nature in modern liberal economics and, indeed, the general claim to scientific precision and impartiality for which modern economists by and large have striven, for better or worse.

In modern economics the dividing line between mainstream economists and their more environmentally inclined peers has become manifest in the different perspectives of environmental versus ecological economics. Mainstream environmental economics attempts to maintain the use of natural resources to the furthest extent possible. While it accepts the finiteness of these resources, it does not see this as a reason to discontinue their utilization. This is evident in the concept of weak sustainability, the idea that a natural resource might be utilized to exhaustion, so long as sufficient investment in other economically equivalent resources or capital is made, so that future generations will possess complementary economic resources. Animals in this sense can be evaluated as equivalent to inanimate resources in a sense not much different from that of Walras. The weak sustainability argument was developed by economists such as Robert Solow, who in fact provides a sober vision of this argument far from disdainful of environmental concerns, a caution not always evident among other proponents of a similar economic philosophy.[95] The different approach of strong sustainability common among ecological economists, even though it remains unpopular among mainstream economists, claims in contrast that some natural resources are incommensurable and cannot be compensated by other resources or by capital.[96]

The divide between economists regarding environmental concerns, including animals, is not always defined along these sustainability lines and has found its way, if inconsistently, into the mainstream of economic debate. Though not a specific advocate of weak sustainability, one of the most original and sophisticated modern advocates for the economic use of nature was Robert Fogel. He was a leading proponent of the modern study of cliometrics, the application of quantitative economic methodology to

95 Solow, "Intergenerational Allocation"; Solow, "Economics of Resources"; Solow, "Does Growth Have a Future?"

96 On these concepts, see Ayres et al., "Strong versus Weak Sustainability"; Bergh, "Ecological Economics"; and Munda, "Environmental Economics."

the study of economic history. Fogel presented one of the most updated advocacies for maximizing the use of natural resources. He developed the concept of "technophysio evolution," the complicated interaction between modern progress in the technology of production which has vastly increased human control of the environment, and improvements in human physiology and longevity. His approach evinces a firm belief that "human beings have gained an unprecedented degree of control over their environment – a degree of control so great that it sets them apart not only from all other species, but also from all previous generations of *Homo Sapiens*."[97] For Fogel this was clearly a positive development. Here was a seemingly novel historical and economic argument for separating humans and animals, and justifying the use of the latter by the former. But as we have seen, despite its novelty, this type of argumentation hails back to the late Enlightenment.

By and large, mainstream modern economic thought has remained insistent in its advocacy for utilizing nature and animals. The need to balance this with social, ethical, and environmental concerns – which, as we saw, was already evident, if only slightly, in Adam Smith's thought – nevertheless occasionally resurfaces, for example, recently, in some of Amartya Sen's writings. Though not an ecological economist, and while not objecting to the basic right of humans to utilize nature, Sen has nonetheless advocated for a more socially responsible use of natural resources, which would result not only in destructive consequences but in ameliorative ones as well, noting that "Development is fundamentally an empowering process, and this power can be used to preserve and enrich the environment, and not only to decimate it. We must not, therefore, think of the environment exclusively in terms of conserving pre-existing natural conditions, since the environment can also include the results of human creation."[98] The latter point is telling, testifying to an attempt to balance utilization of nature with its preservation. But in essence, Sen's outlook is as reminiscent of the Marxian view of nature as of Mill's, since he seems more concerned with presenting a socially just system of natural resource allocation than with conserving the environment for its own sake. Yet even Mill, the truly original and candid economist and philosopher to face this dilemma, eventually, as we have seen, accepted the fact that between the ineluctable need to utilize animals and other natural resources and the pressing need to preserve them, the former ultimately overweighed the latter. This was a sad conclusion, but it was inescapable. Mill's dilemma lives on.

97 Fogel, *Escape from Hunger*, xv, 20-2 (21 for the quotation), 33, 129-30n, and passim.

98 Sen, "Why We Should Preserve"; Sen, "Sustainable Development"; and, for the quotation, Sen, *Idea of Justice*, 248-9.

An Equine Coda

Animals continued to figure prominently in eighteenth-century painting, often based on seventeenth-century Netherlandish models, in the work of artists such as the French painter Jean-Baptiste Oudry and the English painter George Stubbs. Animals also figured prominently in the art of Thomas Gainsborough, and another English painter, William Hogarth, though not an animal painter in the full sense, gave the philosophical topic of attitudes toward animals one of its most famous expressions in his series of engravings *The Four Stages of Cruelty*, even if he was ultimately concerned mainly with the influence this had in forming attitudes toward human beings.[99] For our purposes, however, it is more instructive to see how animal imagery changed in the transition to nineteenth-century romantic art. A short overview of equine paintings can demonstrate the same dialectical tension between economic utilization and aesthetic contemplation which was so significant for Mill and can therefore round off our discussion in a proper manner, bridging intellectual history and art history.

Horses, of course, were the most ubiquitous animals in western art from antiquity to the invention of motorized vehicles. In fact, they still continued to play a vital economic role in nineteenth-century industrialization and transportation.[100] During the early modern era this role had been even more vital, and it is not surprising that modern scholarship has accorded consistent attention to the place of horses in the economic and cultural life of early modern Europe. The role of horses in human economic and social existence was emphatic, whether as beasts of burden, as cavalry horses, or as the animals which carried the élite during their peregrinations through the countryside.[101] Equestrian portraits were customarily connected to symbols of chivalry, defense of faith, regality and the ability to rule, and as such they figured prominently in seventeenth-century baroque art, in paintings by Velázquez, Rubens, van Dyck, and others.[102] In the bourgeois Dutch Republic, however, monumental equestrian portraiture became less common than in other contemporary monarchical countries, and most paintings of this

99 For a recent discussion, see Beirne, "Hogarth's Animals." On Gainsborough, see Cohen, "Thomas Gainsborough's Sensible Animals."
100 McShane and Tarr, "Horse in the Nineteenth-Century."
101 E.g., among many studies, Thirsk, "Horses"; Raber and Tucker, eds., *Culture of the Horse*; Cuneo, "(Un)stable Identities"; Edwards, "Nature Bridled"; and Graham, "Reading, Writing."
102 Liedtke, *Royal Horse*.

Figure 19 Paulus Potter, *The Farrier's Shop*, 1648

Courtesy National Gallery of Art, Widener Collection, Washington D.C.

Figure 20 Paulus Potter, *An Old Horse and a Dead Horse*, 1652

Rijksmuseum, Amsterdam

type were in a rather small format.[103] Dutch equestrian portraits could on occasion symbolize not only social rank but also the enjoyment of country life, and in the young republic the figure of the horse could also symbolize liberty.[104] Yet it was the *naer het leven* view of the place of horses in everyday life, standing at the entrance to roadside inns, being ridden by all social strata, or treated for various purposes as in Potter's painting *The Farrier's Shop* (Fig. 19) which predominated. Potter could contemplate in the most naturalistic fashion the realities of equine life and death, as he did, for example, in his series of five etchings of horses.[105] The one reproduced here (Fig. 20) gives ample evidence that Potter was well aware of the fate that awaited those horses decimated by the use they endured at the hands of their human handlers. Whether he intended this as criticism of the economic use of animals remains doubtful, however. That seems more in line with a modern interpretation of this etching. For him, it was probably part of the customary contemporary *naer het leven* observance of the world. Yet this was a particularly poignant example of how this observance for Potter, as for the other great artists of the Dutch Golden Age, constituted much more than physical exactitude, and also included a piercing psychological insight. *Naer het leven* is therefore a more apt and expansive term in discussing Dutch painting than the more common reference to realism. These etchings by Potter, in any case, were totally devoid of the explicit symbolism of previous depictions of horses in graphic art, most notably Hans Baldung Grien's sixteenth-century woodcut series *Wild Horses*, in which equine sexual desire was presented as an emblematic image of unsettling wild disorder.[106] In his customary manner, Potter, with his more modern outlook, accorded more attention to the animals themselves than to their traditional symbolic significance for humans. He had the uncanny ability to invest horses, like other animals, with expressive facial and psychological content akin to human portraiture (Figs. 21, 22).

103 Exceptions, specifically large-scale equestrian portraits, include Rembrandt's *Portrait of Frederik Rihel(?) on Horseback*, in the National Gallery, London, his famous *The Polish Rider* in the Frick Collection, New York, and Potter's *Equestrian Portrait of Dirk Tulp*, in the Six Collection, Amsterdam. The growing demand for equestrian portraits of rich burghers in traditional aristocratic poses was indicative of the changing social conditions in seventeenth-century Holland.
104 For general remarks on Dutch seventeenth-century equestrian art, see Dumas, *In het Zadel*, 15-25, 28. On the horse as a Dutch symbol of liberty, see Walsh, Buijsen, and Broos, *Paulus Potter*, 146.
105 For reproductions and discussion, see Walsh, Buijsen, and Broos, *Paulus Potter*, 194-7; and for other examples also ibid., 176-77, 180.
106 Koerner, *Moment of Self-Portraiture*, 426-37.

Figure 21 Paulus Potter, Detail from *The "Piebald" Horse*, c. 1650-1654

J. Paul Getty Museum, Los Angeles

Figure 22 Paulus Potter, Detail from *A Spaniel* (fig. 11), 1653

Rijksmuseum, Amsterdam

The depiction of horses, as of other animals, in seventeenth-century art mirrored the receding of traditional symbolism in favor of a more straightforward observation of the animals' life, both in themselves and in relation to their place in everyday human life, not least from an economic perspective. Here in a nutshell was the early tension between the significant practical place that utilizing animals played in economic and social existence and the tendency to idealize them. The latter option was now less inclined to traditional symbolism and allegory and more toward the idealization of country life and nature which would become so poignant in romantic art. Horses, of course, continued to play a prominent role in eighteenth-century iconography, not unlike that which they had played in seventeenth-century art. This tension between utilization and idealization of nature and animals continued to be felt throughout the seventeenth, eighteenth, and nineteenth centuries. If we jump directly to nineteenth-century romantic art, about the same time that Mill was writing and rewriting his great work on political economy, we encounter this ambiguity in full force. Eugène Delacroix's painting *Arab Horses Fighting in a Stable* is a striking example, giving full vent to the artist's excitement at the physical and emotional drama of animal combat (Fig. 23). The painting was based on a real incident which the painter witnessed on his formative North African journey and recorded in detail in his journal, where he described how the horses "reared up and fought with a fury that made me tremble for their riders, but magnificent for painting."[107] This romantic enthusiasm for horses was an even more central concern for his contemporary Théodore Géricault, probably the most consistent painter of equine motifs among romantic artists.[108] Géricault, however, also scrutinized in detail the more mundane aspects of horse life. His painting *The Plaster Kiln* is an example (Fig. 24). The overcast sky conveys a romantic sense of melancholy. Are the horses there only to transport industrial material, or are they themselves about to be turned into such material? In any case, these are not noble Arabian horses fighting but, rather, ordinary workhorses leading as dull a life as their laboring human owners. It was this reality which contemporaries would have associated most often with everyday experience of horses, whose economic utilization continued unabated, and perhaps even intensified, in the era of early industrialization. The contrast, or rather coexistence, of these two equally romantic equine images by Delacroix and Géricault mirrors the same duality of attitudes toward animals which we have constantly encountered. This duality was

107 Delacroix, *Journal*, 52.
108 Grunchec, *Géricault's Horses*.

Figure 23 Eugène Delacroix, *Arab Horses Fighting in a Stable*, 1860

Musée d'Orsay, Paris

a sign of the complicated relationship of modern urban civilization with nature. It became more poignant the more urban and industrialized a society was; hence it was a sign of modernity, of the modern conceptualization of animals. This is perhaps one of the main reasons that this new perception of animals developed so early in Holland, which in the seventeenth century was the most socially and economically advanced and stable country in Europe and the one most closely approximating what would become modern bourgeois democratic civilization. The historical development which led to this situation in the nineteenth century was, as we have seen, the result of a process going back to the seventeenth century. In this as in so many other respects, the long eighteenth century was the the transitional period between the early modern and the modern eras, and it shared characteristics of both. How animals were viewed and discussed in the early seventeenth century compared with two centuries later might seem to suggest a deep divide. But once we recognize the changes that happened between these two points, the connecting historical line between them seems coherent and logical.

Figure 24 Théodore Géricault, *The Plaster Kiln*, 1822-1823

Musée du Louvre, Paris

Yet again we find seventeenth-century Dutch art mirroring this whole change already at an early stage of its development, and providing an image of the growing reflexive awareness in European culture of the specifically economic role that animals played in human social existence. Gerard ter Borch's arresting image *Man on Horseback* (Fig. 25) conveys the transition in attitudes toward horses, and indeed toward all animals, from the traditional symbolic perspective to the more quotidian down-to-earth outlook of nascent modernity, with its distinctly economic outlook.[109] Ter Borch's rider, perhaps a freshly unemployed soldier returning home, and his horse, turn their backs on what perhaps was a glorious personal military past. They slump forward, rather reluctantly, seeming to accept the unavoidable drab fate awaiting them, one of mundane work and subsistence. For the animal, at least, this is a future fraught with as many dangers and difficulties as the past, if of a less illustrious nature.

109 Ter Borch painted at least one other version of this composition. Also see Wheelock et al., *Gerard ter Borch*, 44-6; and Schama, *Embarrassment of Riches*, 243-4.

Figure 25 Gerard ter Borch, *Man on Horseback*, 1634

Museum of Fine Arts, Boston

Bibliography

Primary Sources

Adair, James, *The History of the American Indians*, ed. by Kathryn E. Holland Braund (Tuscaloosa: The University of Alabama Press, 2005).

Addison, Joseph, *The Spectator*, ed. by Gregory Smith, 4 vols. (London and New York: Dent, 1966-67).

Ammianus Marcellinus, [*Works*], trans. by John C. Rolfe, 3 vols. (Cambridge, MA: Harvard University Press, Loeb Classical Library, 1935-39).

Aquinas, St. Thomas, *Political Writings*, ed. and trans. by R. W. Dyson (Cambridge: Cambridge University Press, 2002).

Aristotle, *The Politics*, trans. by Benjamin Jowett, revised by Jonathan Barnes, ed. by Stephen Everson (Cambridge: Cambridge University Press, 1988).

Beattie, James, *Dissertations Moral and Political* (London: W. Strahan and T. Cadell; and Edinburgh: W. Creech, 1783; repr. London: Routledge, 1996).

Benesch, Otto, enlarged and edited by Eva Benesch, *The Drawings of Rembrandt*, 6 vols. (London: Phaidon, 1973).

Bentham, Jeremy, *An Introduction to the Principles of Morals and Legislation*, ed. by J. H. Burns and H. L. A. Hart (London: University of London, 1970).

—, "Letter to the editor of *The Morning Chronicle* from 4 March 1825," in *Political Theory and Animal Rights*, ed. by Paul A. B. Clarke and Andrew Linzey (London and Winchester: Pluto Press, 1990), 135-7.

Bie, Cornelis de, *Het Gulden Cabinet van de Edel Vry Schilderconst* (Antwerp: Ian Meyssens, 1662; repr. Soest: Davaco, 1971).

Boswell, James, *Boswell's Life of Samuel Johnson*, abridged by Anne and Irvin Ehrenpreis (New York: Washington Square Press, 1965).

Buffon, Georges-Louis Leclerc, Comte de, *Histoire naturelle, générale et particulière*, 36 vols. (Paris: Imprimerie Royale, 1749-89).

—, *Natural History, General and Particular*, trans. by William Smellie, 9 vols., 3rd edn. (London: A. Strahan and T. Cadell, 1791; repr. Bristol: Thoemmes Press, 2000).

Cato and Varro, *On Agriculture*, trans. W. D. Hooper and Harrison Boyd Ash (Cambridge, MA: Harvard University Press, Loeb Classical Library, 1934).

Cats, Jacob, *Zinne-en Minne-Beelden* (Amsterdam: de Wed. Barend Visscher, 1720 [1618]).

Cavendish, Margaret, the Duchess of Newcastle, *Poems, or, Several Fancies in Verse: with the Animal Parliament, in Prose* [3rd ed.] (London: A. Maxwell, 1668).

Cervantes, Miguel de, *The Adventures of Don Quixote*, trans. by J. M. Cohen (Harmondsworth: Penguin Books, 1952).

Clavigero, Francisco, *The History of Mexico*, trans. by Charles Cullen, 2 vols. (London: G. G. J. and J. Robinson, 1787; repr. New York and London: Garland Publishing, 1979).

Court, Pieter de la, *The True Interest and Political Maxims, Of the Republic of Holland*, trans. by John Campbell (London: J. Nourse, 1746).

Crèvecoeur, J. Hector St. John de, *Letters from an American Farmer and Sketches of Eighteenth-Century America*, ed. by Albert E. Stone (Harmondsworth: Penguin Books, 1981).

Delacroix, Eugène, *The Journal of Eugène Delacroix*, ed. by Hubert Wellington, trans. by Lucy Norton (London: Phaidon, 1951).

Descartes, René, *Oeuvres de Descartes*, ed. by Charles Adam and Paul Tannery, 11 vols. (Paris: Librairie philosophique J. Vrin, 1996).

—, *The Philosophical Works of Descartes*, trans. by Elizabeth S. Haldane and G. R. T. Ross, 2 vols. (Cambridge: Cambridge University Press, 1967).
Dunbar, James, *Essays on the History of Mankind in Rude and Cultivated Ages* (London: W. Strahan, 1781; repr. Bristol: Thoemmes Press, 1995).
The Eight Volumes of Letters Writ by a Turkish Spy, who lived five and forty years undiscover'd, at Paris: giving an impartial account to the Divan at Constantinople, of the most remarkable transactions of Europe, and discovering several intrigues and secrets of the Christian courts (especially that of France) from the year 1637 to the year 1682, 8 vols. (London: H. Rhodes et al., 1694).
Engels, Frederick, *Dialectics of Nature*, trans. by Clemens Dutt (Moscow: Foreign Languages Publishing House, 1954).
Evelyn, John, *The Diary of John Evelyn*, ed. by John Bowle (Oxford: Oxford University Press, 1985).
Ferguson, Adam, *An Essay on the History of Civil Society*, ed. by Fania Oz-Salzberger (Cambridge: Cambridge University Press, 1995).
—, *Principles of Moral and Political Science*, 2 vols. (Edinburgh: W. Creech, 1792; repr. Hildesheim and New York: G. Olms, 1975).
Fogel, Robert William, *The Escape from Hunger and Premature Death, 1700-2100: Europe, America, and the Third World* (Cambridge: Cambridge University Press, 2004).
Gibbon, Edward, *The English Essays of Edward Gibbon*, ed. by Patricia B. Craddock (Oxford: Clarendon Press, 1972).
—, *Essai sur l'étude de la littérature and An Essay on the Study of Literature*, with Introduction by John Valdimir Price (Dublin: Guillaume Hallhead, 1777 [the French edition] and London: T. Becket and P. A. De Hondt, 1764 [the English translation]; repr. London: Routledge and Thoemmes Press, 1994).
—, *Gibbon's Journal to January 28th, 1763, My Journal, I, II & III and Ephemerides*, ed. by D. M. Low (London: Chatto & Windus, 1929).
—, *The History of the Decline and Fall of the Roman Empire*, ed. by David Womersley, 3 vols. (Harmondsworth: Penguin Books, 1995).
—, *The Letters of Edward Gibbon*, ed. by J. E. Norton, 3 vols. (London: Cassel, 1956).
Godefroy, Louis, *The Complete Etchings of Adriaen van Ostade* (San Francisco: Alan Wofsy Fine Arts, 1990).
Goguet, Antoine-Yves, *The Origin of Laws, Arts, and Sciences, and Their Progress Among the Most Ancient Nations*, trans. Robert Henry[?], 3 vols. (Edinburgh: Alex Donaldson and John Reid, 1761; repr. New York: AMS Press, 1976).
Goldsmith, Oliver, *The Works of Oliver Goldsmith* (Edinburgh: William P. Nimmo, 1870?).
Graaf, Regnier de, "Regnier de Graaf on the Human Reproductive Organs," trans. by H. D. Jocelyn and B. P. Setchell, *Journal of Reproduction and Fertility*, Supplement 17 (1972), 1-222.
Gregory, John, *A Comparative View of the State and Faculties of Man with Those of the Animal World*, 6th ed., 2 vols. [1 in the modern reprint] (London: J. Dodsley, 1774; repr. Tokyo: Thoemmes Press, 1994).
—, *A Father's Legacy to His Daughters* (London: W. Strahan and T. Cadell, and Edinburgh: W. Creech, 1774; repr. New York and London: Garland Publishing, 1974).
—, *John Gregory's Writings on Medical Ethics and Philosophy of Medicine*, ed. by Laurence B. McCullough (Dordrecht: Kluwer Academic Publishers, 1998).
Grunchec, Philippe, *Géricault's Horses: Drawings and Watercolours* (London: Sotheby's Publications, 1985).
Guericke, Otto von, *The New (So-Called) Magdeburg Experiments of Otto von Guericke*, trans. by Margaret G. F. Ames (Dordrecht: Kluwer, 1994).
Harvey, William, *The Circulation of the Blood and Other Writings*, trans. by Kenneth J. Franklin (London: Dent, 1966).

Hemingway, Ernest, *Green Hills of Africa* (New York: Permabooks, 1954).
Henry, Robert, *The History of Great Britain, from the Invasion of it by the Romans Under Julius Cæsar, Written on a New Plan*, 6 vols. (London: T. Cadell [vols. 4-5 also T. Longman, vol. 6 also A. Strahan, vol. 5 also in Edinburgh], 1771-93).
Herder, Johann Gottfried, *Outlines of a Philosophy of the History of Man*, trans. by T. Churchill (London: J. Johnson, 1800).
Home, Henry (Lord Kames), *Sketches of the History of Man*, 4 vols. (Edinburgh: W. Creech, 1778; repr. London: Routledge, 1993).
Hoogstraeten, Samuel van, *Inleiding tot de Hooge Schoole der Schilderkonst* (Rotterdam: Fransois van Hoogstratten, 1678; repr. n.p.: Davaco, 1969).
Houbraken, Arnold, *De Groote Schouburgh der Nederlantsche Konstschilders en Schilderessen*, 3 vols. (Amsterdam: Arnold Houbraken, 1718-21).
Hume, David, *The History of England, from the Invasion of Julius Caesar to the Revolution in 1688*, 6 vols. (Indianapolis: Liberty Classics, 1983).
—, *An Inquiry Concerning Human Understanding*, ed. by Tom L. Beauchamp (Oxford: Oxford University Press, 1999).
—, *A Treatise of Human Nature*, ed. by L. A. Selby-Bigge (Oxford: Clarendon Press, 1967).
Hutcheson, Francis, *A System of Moral Philosophy*, 2 vols. (London: A. Millar, 1755; repr. Bristol and Sterling: Thoemmes Press, 2000).
Hutton, James, *An Investigation of the Principles of Knowledge, and of the Progress of Reason, from Sense to Science and Philosophy*, 3 vols. (Edinburgh and London: A. Strahan and T. Cadell, 1794).
—, *Theory of the Earth, with Proofs and Illustrations*, 2 vols. (London: Cadell, Junior, and Davies, and Edinburgh: William Creech, 1795).
Huygens, Christianus [Christiaan], *The Celestial Worlds Discover'd: Or, Conjectures Concerning the Inhabitants, Plants and Productions of the Worlds in the Planets*, trans. by Anon. (London: Timothy Childe, 1698).
—, *Oeuvres Complètes*, 22 vols. (The Hague: Martinus Nijhoff, 1888-1950).
Johnson, Samuel, *The Idler and The Adventurer*, ed. by W. J. Bate, John M. Bullitt, and L. F. Powell (New Haven and London: Yale University Press, 1963).
Kant, Immanuel, *Lectures on Ethics*, trans. Louis Infield (New York: Harper & Row, 1963).
Kerr, Robert, *Memoirs of the Life, Writings, and Correspondence of William Smellie*, Introduction by Richard B. Sher, 2 vols. (Edinburgh: John Anderson, and London: Longman et al., 1811; repr. Bristol: Thoemmes Press, 1996).
Laet, Joannes de, *The Empire of the Great Mogol*, trans. by J. S. Hoyland (New Delhi: Oriental Books Reprint / Munshiram Manoharlal, 1974).
Lairesse, Gerard de, *The Art of Painting*, trans. by John F. Fritsch (London: S. Vandenbergh et al., 1778).
Leibniz, Gottfried Wilhelm, *Philosophical Papers and Letters*, trans. by L. E. Loemker (Dordrecht: D. Reidel, 1969).
Letters Writ by a Turkish Spy, ed. by Arthur J. Weitzman (London: Routledge and Kegan Paul, 1970).
Lewis, Rhodri, *William Petty on the Order of Nature: An Unpublished Manuscript Treatise* (Tempe: Arizona Center for Medieval and Renaissance Studies, 2012).
Linschoten, Jan Huygen van, *Discourse of Voyages into Y East and West Indies*, trans. by W. P. [*sic*] (London: Iohn Wolfe, 1598; repr. Amsterdam and Norwood: Theatrum Orbis Terrarum, 1974).
Lipsius, Justus, *A Discourse of Constancy*, trans. by R. G. [*sic*] (London: Humphrey Moseley, 1654).
Luiken, Johannes [Jan], and Caspaares [Kasper] Luiken, *Het Menselyk Bedryf* (Amsterdam: Johannes and Caspaares Luiken, 1694).
Malthus, T. R., *An Essay on the Principle of Population*, ed. by Patricia James, 2 vols. (Cambridge: Cambridge University Press, 1989).

Mander, Carel [Karel] van, *Het Schilder-Boeck* (Haarlem: Paschier van Wesbusch, 1604; repr. Utrecht: Davaco, 1969).
McCulloch, J. R., ed., *Early English Tracts on Commerce* (Cambridge: Cambridge University Press, 1970 [1856]).
Meek, Ronald L., *The Economics of Physiocracy: Essays and Translations* (Cambridge, MA: Harvard University Press, 1963).
La Mettrie, Julien Offray de, *Machine Man and Other Writings*, trans. by Ann Thomson (Cambridge: Cambridge University Press, 1996).
Mill, John Stuart, *Essays on Ethics, Religion and Society*, ed. by J. M. Robson (Toronto: University of Toronto Press, 1969).
—, *Principles of Political Economy, with Some of Their Applications to Social Philosophy*, ed. by J. M. Robson, 2 vols. (Toronto: University of Toronto Press, 1965).
Montaigne, Michel de, *The Complete Essays*, trans. by M. A. Screech (Harmondsworth: Penguin Books, 1993).
Montesquieu, *Considerations on the Causes of the Greatness of the Romans and their Decline*, trans. and ed. by David Lowenthal (New York and London: The Free Press and Collier-Macmillan, 1965).
—, *Persian Letters*, trans. by C. J. Betts (Harmondsworth: Penguin Books, 1973).
—, *The Spirit of the Laws*, trans. and ed. by Anne M. Cohler, Basia Carolyn Miller, and Harold Samuel Stone (Cambridge: Cambridge University Press, 1990).
Nieuhoff, Jan, *An Embassy from the East-India Company of the United Provinces, to the Grand Tartar Cham Emperour of China*, trans. by John Ogilby (London: John Macock, 1669; repr. Menston: Scolar Press, n.d.).
Oswald, John, *The Cry of Nature; Or, An Appeal to Mercy and to Justice on Behalf of the Persecuted Animals (1791)*, ed. by Jason Hribal (Lewiston: Edwin Mellen Press, 2000).
Ovid, *Metamorphoses*, trans. by A.D. Melville (Oxford and New York: Oxford University Press, 1992).
Parsons, Howard L., ed., *Marx and Engels on Ecology* (Westport, CT: Greenwood Press, 1977).
Passe, C[h]rispijn van de [the Younger], *'t Light der Teken en Schilderkonst* (Amsterdam: I. Iantz., 1643; repr. Soest: Davaco, 1973).
Petty, William, *Several Essays in Political Arithmetick* (London, D. Browne et al., 1755; repr. London: Routledge and Thoemmes Press, and Tokyo: Kinokuniya, 1992).
Pope, Alexander, *The Works of Alexander Pope*, 10 vols. (New York: Gordian, 1967).
Pope, Alexander, John Arbuthnot, Jonathan Swift, John Gay, Thomas Parnell, and Robert Harley, *The Memoirs of the Extraordinary Life, Works, and Discoveries of Martinus Scriblerus*, ed. by Charles Kerby-Miller (New York and Oxford: Oxford University Press, 1988).
Porphyry, *On Abstinence from Animal Food*, trans. by Thomas Taylor (London and Fontwell: Centaur Press, 1965).
Primatt, Humphry, *The Duty of Mercy and the Sin of Cruelty to Brute Animals*, ed. by Richard D. Ryder (Fontwell: Centaur Press, 1992 [1776]).
Procopius, [*Works*], trans. by H. B. Dewing, 7 vols. (Loeb Classical Library, 1914-40).
Rabelais, François, *The Complete Works of François Rabelais*, trans. by Donald M. Frame (Berkeley: University of California Press, 1991).
Raynal, Guillaume-Thomas, Abbé de, *Philosophical and Political History of the Settlements and Trade of the Europeans in the East and West Indies ... By the Abbé Raynal*, trans. by J. O. Justamond, 6 vols., 2nd ed. (London: J. Mundell & Co., 1798; repr. New York: Negro Universities Press, 1969).
Redi, Francesco, *Francesco Redi on Vipers*, trans. by Peter K. Knoefel (Leiden: Brill, 1988).

Reid, Thomas, *Practical Ethics, Being Lectures and Papers on Natural Religion, Self-Government, Natural Jurisprudence, and the Law of Nations*, ed. by Knud Haakonssen (Princeton: Princeton University Press, 1990 [1751-96]).

—, *The Works of Thomas Reid*, ed. by William Hamilton, 2 vols. (Edinburgh: MacLachlan and Stewart, and London: Longman et al., 1863; repr. Bristol: Thoemmes Press, 1994).

Mirabeau, Victor de Riqueti, Marquis de, *Philosophie rurale, ou économie générale et politique de l'agriculture, reduite à l'ordre immuable des lois physiques et morales, qui assurent la prospérité des empires*, 3 vols. (Amsterdam: Libraires Associés, 1764).

Robertson, William, *An Historical Disquisition Concerning the Knowledge Which the Ancients Had of India* (London: A. Strahan and T. Cadell, and Edinburgh: E. Balfour, 1794; repr. London: Routledge/Thoemmes Press, 1996).

—, *The History of America*, 6th ed., 3 vols. (London: A. Strahan and T. Cadell, and Edinburgh: J. Balfour, 1792; repr. London: Routledge/Thoemmes Press, 1996).

Rousseau, Jean-Jacques, *Correspondance générale de J.-J. Rousseau*, ed. by Théophile Dufour (Paris: Librairie Armand Colin, 1924).

—, *Oeuvres complètes*, ed. by Bernard Gagnebin and Marcel Raymond, 5 vols. (Paris: Gallimard, 1959-95).

Ruskin, John, *Modern Painters*, 5 vols. (New York: Clarke, Given & Hooper, 1873).

Salt, Henry S., *Animals' Rights Considered in Relation to Social Progress* (New York: Macmillan & Co., 1894).

Say, Jean-Baptiste, *A Treatise on Political Economy, Or the Production, Distribution and Consumption of Wealth*, trans. by C. R. Prinsep, Introduction trans. by Clement C. Biddle (Philadelphia: Claxton, Remsen & Haffelfinger, 1880; repr. New York: Augustus M. Kelley, 1971).

Schwartz, Gary, ed., *The Complete Etchings of Rembrandt* (New York: Dover, 1994).

Sen, Amartya, *The Idea of Justice* (London: Penguin Books, 2010).

—, "Sustainable Development and Our Responsibilities," *Notizie di Politeia*, 26 (2010), 129-37.

—, "Why We Should Preserve the Spotted Owl," *London Review of Books*, 26 (5 February, 2004), 10-11.

Simonde de Sismondi, J.-C.-L., *New Principles of Political Economy: Of Wealth in Its Relation to Population*, trans. by Richard Hyse (New Brunswick and London: Transaction Publishers, 1991).

Smellie, William, *Literary and Characteristical Lives of John Gregory, M.D., Henry Home, Lord Kames, David Hume, Esq., and Adam Smith, L.L.D., to which are added a Dissertation on Public Spirit; and Three Essays*, Introduction by Stephen Brown (Edinburgh and London: Alexander Smellie et al., 1800; repr. Bristol: Thoemmes Press, 1997).

—, *The Philosophy of Natural History*, 2 vols. (Edinburgh and London: various publishers, 1790, 1799; repr. Bristol and Sterling: Thoemmes Press, 2001).

Smith, Adam, *Essays on Philosophical Subjects*, ed. by W. P. D. Wightman (Oxford: Clarendon Press, 1980).

—, *An Inquiry into the Nature and Causes of the Wealth of Nations*, ed. by R. H. Campbell, A. S. Skinner, and W. B. Todd, 2 vols. (Oxford: Clarendon Press, 1979).

—, *Lectures on Jurisprudence*, ed. by R. L. Meek, D. D. Raphael, and P. G. Stein (Oxford: Clarendon Press, 1978).

—, *Lectures on Rhetoric and Belles Lettres*, ed. by J. C. Bryce (Oxford: Clarendon Press, 1983).

—, *The Theory of Moral Sentiments*, ed. by D. D. Raphael and A. L. Macfie (Oxford: Clarendon Press, 1976).

Solow, Robert M., "Does Growth Have a Future? Does Growth Theory Have a Future? Are These Questions Related?" *History of Political Economy*, 41, annual supplement (2009), 27-34.

—, "The Economics of Resources or the Resources of Economics," *American Economic Review*, 64 (1974), 1-14.

—, "On the Intergenerational Allocation of Natural Resources," *Scandinavian Journal of Economics*, 88 (1986), 141-9.
Steele, Richard, *The Tatler* (any edition).
Stewart, Dugald, *Outlines of Moral Philosophy* (Edinburgh: W. Creech, 1793; repr. New York and London: Garland, 1976).
Swammerdam, Jan, *The Letters of Jan Swammerdam to Melchisedec Thévenot*, ed. by G. A. Lindeboom (Amsterdam: Swets & Zeitlinger, 1975).
Taylor, Helen, "A Few Words on Mr. Trollope's Defence of Fox-Hunting," *Fortnightly Review*, 37 (January 1870), 63-8.
Thucydides, *History of the Peloponnesian War*, trans. by Rex Warner (Harmondsworth: Penguin Books, 1972).
Transactions of the Royal Society of Edinburgh, vol. 1 (Edinburgh: J. Dickson, 1788).
Turgenev, Ivan, "Memoirs of a Sportsman," in *Memoirs of a Sportsman and A Nobleman's Nest*, trans. by Isabel F. Hapgood (Boston and New York: Jefferson Press, 1903).
Ulman, H. Lewis, ed., *The Minutes of the Aberdeen Philosophical Society 1758-1773* (Aberdeen: Aberdeen University Press, 1990).
Uztáriz, Jerónimo de, *The Theory and Practice of Commerce and Maritime Affairs*, trans. by John Kippax (Dublin: George Faulkner, 1752).
Vesalius, Andreas, "The Last Chapter of the 'De Fabrica' of Vesalius, Entitled Some Observations on the Dissection of Living Animals," trans. by Benjamin Farrington, *Transactions of the Royal Society of South Africa*, 20 (1932), 1-14.
Vico, Giambattista, *New Science: Principles of the New Science Concerning the Common Nature of Nations, Third Edition*, trans. by David Marsh (Harmondsworth: Penguin Books, 1999).
Visscher, Roemer, *Sinnepoppen* (Amsterdam: Willem Iansz., 1614).
Voltaire, "Dialogue between the Cock and the Hen," trans. by Haydn and Adrienne Mason, *Comparative Criticism*, 20 (1998), 183-7.
—, *Dictionnaire philosophique*, ed. by Christiane Mervaud, in *The Complete Works of Voltaire*, vol. 1 (Oxford: Voltaire Foundation, 1994).
—, *Essai sur les mœurs et l'esprit des nations et sur les principaux faits de l'histoire depuis Charlemagne jusqu'à Louis XIII*, ed. by René Pomeau, 2 vols. (Paris: Éditions Garnier Frères, 1963).
Walras, Léon, *Elements of Pure Economics, or the Theory of Social Wealth*, trans. by William Jaffé (Homewood, IL: Richard D. Irwin, 1954).
Young, Thomas, *An Essay on Humanity to Animals (1798)*, ed. by Rod Preece (Lewiston: Edwin Mellen Press, 2001).
Zambeccari, Joseph [Giuseppe], "Experiments of Doctor Joseph Zambeccari Concerning the Excision of Various Organs from Different Living Animals," trans. by Saul Jarcho, *Bulletin of the History of Medicine*, 9 (1941), 311-25.

Secondary Sources

Allen, Gay Wilson, and Roger Asselineau, *St. John de Crèvecoeur: The Life of an American Farmer* (New York: Viking, 1987).
Alpers, Svetlana, *The Art of Describing: Dutch Art in the Seventeenth Century* (London: Penguin Books, 1983).
—, "Picturing Dutch Culture," in *Looking at Seventeenth-Century Dutch Art: Realism Reconsidered*, ed. by Wayne Franits (Cambridge: Cambridge University Press, 1997), 57-67.

Arnold, Denis G., "Hume on the Moral Difference between Humans and Other Animals," *History of Philosophy Quarterly*, 12 (1995), 303-16.

Ault, Norman, "Pope and his Dogs," in *New Light on Pope* (Hamden: Archon, 1967), 337-50.

Ayres, Robert U., Jeroen C.J.M. van den Bergh, and John M. Gowdy, "Strong versus Weak Sustainability: Economics, Natural Sciences, and 'Consilience'," *Environmental Ethics*, 23 (2001), 155-68.

Baier, Annette C., "Knowing Our Place in the Animal World," in *Ethics and Animals*, ed. by Harlan B. Miller and William H. Williams (Clifton: Humana Press, 1983), 61-77.

Baker, Steve, *Picturing the Beast: Animals, Identity and Representation* (Manchester and New York: Manchester University Press, 1993).

—, "'You Kill Things to Look at Them': Animal Death in Contemporary Art," in The Animal Studies Group, *Killing Animals* (Urbana and Chicago: University of Illinois Press, 2006), 6-98.

Bal, Mieke, *Reading Rembrandt: Beyond the Word-Image Opposition* (Cambridge: Cambridge University Press, 1991).

Balis, Arnout, "Facetten van de Vlaamse Dierenschilderkunst van de 15de tot de 17de Eeuw," in *Het Aards Paradijs: Dierenvoorstellingen in de Nederlanden van de 16de en 17de eeuw*, exhibition catalogue (Antwerpen: Koninklijke Maatschappij voor Dierkunde van Antwerpen and Standaard Uitgeverij, 1982), 37-55.

—, *Rubens' Hunting Scenes [Part XVIII, II, in Corpus Rubenianum Ludwig Burchard]*, trans. by P.S. Falla (Oxford and New York: Harvey Miller, 1986).

Balz, Albert G. A., "Cartesian Doctrine and the Animal Soul," in *Cartesian Studies* (New York: Columbia University Press, 1951), 106-57.

Banzhaf, H. Spencer, "Productive Nature and the Net Product: Quesnay's Economies Animal and Political," *History of Political Economy*, 32 (2000), 517-51.

Barkdull, John, "How Green is the Theory of Moral Sentiments?" in *Land, Value, Community: Callicott and Environmental Philosophy*, ed. by Wayne Ouderkirk and Jim Hill (Albany: State University of New York Press, 2002), 37-58.

Bayertz, Kurt, "The Nature of Morality and the Morality of Nature: Problems and Paradoxes of a Normative Natural Philosophy," in *Nature and Society in Historical Context*, ed. by Mikuláš Teich, Roy Porter, and Bo Gustafsson (Cambridge: Cambridge University Press, 1997), 364-81.

Becker, Christian, Malte Faber, Kirsten Hertel, and Reiner Manstetten, "Malthus vs. Wordsworth: Perspectives on Humankind, Nature and Economy. A Contribution to the History and the Foundations of Ecological Economics," *Ecological Economics*, 53 (2005), 299-310.

Bedini, Silvio A., *The Pope's Elephant* (Nashville: J. S. Sanders, 1998).

Beier, Max, "The Early Naturalists and Anatomists During the Renaissance and Seventeenth Century," in *History of Entomology*, ed. by Ray F. Smith, Thomas E. Mittler, and Carroll N. Smith (Palo Alto: Annual Reviews, in Cooperation with the Entomological Society of America, 1973), 81-94.

Beirne, Piers, "Hogarth's Animals," *Journal of Animal Ethics*, 3 (2013), 133-62.

Beirne, Piers, and Janine Janssen, "Hunting Worlds Turned Upside Down: Paulus Potter's Life of a Hunter," *Tijdschrift over Cultuur & Criminaliteit*, 2 (2014), 15-28.

Bell, Ernest A., *Christian Huygens and the Development of Science in the Seventeenth Century* (London: E. Arnold, 1950).

Benton, Ted, "Engels and the Politics of Nature," in *Engels Today: A Centenary Appreciation*, ed. by Christopher J. Arthur (Basingstoke: Macmillan, 1996), 67-93.

Berger, John, "Why Look at Animals?" in *About Looking* (New York: Pantheon Books, 1980), 1-26.

Bergh, Jeroen C.J.M. van den, "Ecological Economics: Themes, Approaches, and Differences with Environmental Economics," *Regional Environmental Change*, 2 (2001), 13-23.

Bergström, Ingvar, *Dutch Still-Life Painting in the Seventeenth Century*, trans. Christina Hedström and Gerald Taylor (New York: Hacker Art Books, 1983).

Berry, Christopher J., "Smith and Science," in *The Cambridge Companion to Adam Smith*, ed. Knud Haakonssen (Cambridge: Cambridge University Press, 2006), 112-35.

—, *Social Theory of the Scottish Enlightenment* (Edinburgh: Edinburgh University Press, 2001).

Bestiaire hollandais, exhibition catalogue (Paris: Institut Néerlandais, 1960).

Bialostocki, Jan, "Les bêtes et les humains de Roelant Savery," *Bulletin Musées Royaux des Beaux-Arts*, 7 (1958), 69-92.

Blankert, A., "Over Pieter van Laer als Dier- en Landschapschilder," *Oud Holland*, 83 (1968), 117-34.

Blum, Carol, *Strength in Numbers: Population, Reproduction, and Power in Eighteenth-Century France* (Baltimore and London: Johns Hopkins University Press, 2002).

Boas, George, *The Happy Beast in French Thought of the Seventeenth Century* (New York: Octagon Books, 1966 [1933]).

Boehrer, Bruce, *Shakespeare Among the Animals: Nature and Society in the Drama of Early Modern England* (New York and Basingstoke: Palgrave, 2002).

Bonar, James, *Theories of Population from Raleigh to Arthur Young* (London: Frank Cass, 1966).

Boonin-Vail, David, "The Vegetarian Savage: Rousseau's Critique of Meat Eating," *Environmental Ethics*, 15 (1993), 75-84.

Bos, H. J. M., "Christiaan Huygens – a Biographical Sketch," in *Studies on Christiaan Huygens: Invited Papers from the Symposium on the Life and Work of Christiaan Huygens, Amsterdam, 22-25 August 1979*, ed. by H. J. M. Bos, M. J. S. Rudwick, H. A. M. Snelders, and R. P. W. Visser (Lisse: Swets & Zeitlinger, 1980), 7-18.

—, "Huygens, Christiaan", *Dictionary of Scientific Biography*, ed. by Charles Coulston Gillispie, 16 vols. in 8 (New York: Charles Scribner's Sons, 1981), 6: 597-613.

Bowler, Richard, "Sentient Nature and Human Economy: the 'Human' Science of Early *Nationalökonomie*," *History of the Human Sciences*, 18 (2005), 23-54.

Bradie, Michael, "The Moral Status of Animals in Eighteenth-Century British Philosophy," in *Biology and the Foundation of Ethics*, ed. by Jane Maienschein and Michael Ruse (Cambridge: Cambridge University Press, 1999), 32-51.

Braunschneider, Theresa, "The Lady and the Lapdog: Mixed Ethnicity in Constantinople, Fashionable Pets in Britain," in *Humans and Other Animals in Eighteenth-Century British Culture: Representation, Hybridity, Ethics*, ed. by Frank Palmeri (Aldershot and Burlington: Ashgate, 2006), 31-48.

Brienen, Rebecca Parker, "From Brazil to Europe: The Zoological Drawings of Albert Eckhout and Georg Marcgraf," in *Early Modern Zoology: The Construction of Animals in Science, Literature and the Visual Arts*, ed. by Karl A. E. Enenkel and Paul J. Smith, 2 vols. (Leiden and Boston: Brill, 2007), 1: 273-314.

Brown, Laura, "The Lady, the Lapdog, and Literary Alterity," *The Eighteenth Century: Theory and Interpretation*, 52 (2011), 31-45.

Brown, Stephen, "Introduction," in William Smellie, *Literary and Characteristical Lives of John Gregory, M.D., Henry Home, Lord Kames, David Hume, Esq., and Adam Smith, L.L.D., to which are added a Dissertation on Public Spirit; and Three Essays* (Edinburgh and London: Alexander Smellie et al., 1800; repr. Bristol: Thoemmes Press, 1997).

—, "William Smellie 1740-95," in *The Dictionary of Eighteenth-Century British Philosophers*, 2 vols. (Bristol: Thoemmes Press, 1999), 2: 804-8.

Brown, Stephen W., "William Smellie and Natural History: Dissent and Dissemination," in *Science and Medicine in the Scottish Enlightenment*, ed. by Charles W. J. Withers and Paul Wood (East Linton: Tuckwell Press, 2002), 191-214.

Brusati, Celeste, "Natural Artifice and Material Values in Dutch Still Life," in *Looking at Seventeenth-Century Dutch Art: Realism Reconsidered*, ed. by Wayne Franits (Cambridge: Cambridge University Press, 1997), 144-57.

Bruyn, J., "A Turning-Point in the History of Dutch Art," in *Dawn of the Golden Age: Northern Netherlandish Art, 1580-1620*, exhibition catalogue, ed. by Ger Luijten, Ariane van Suchtelen, Reinier Baarsen, and Walter Kloek (Amsterdam: Rijksmusem, and Zwolle: Waanders, 1993), 112-21.

Burch, Christopher B., *Christiaan Huygens: The Development of a Scientific Research Program in the Foundations of Mechanics* (Ph.D. diss., University of Pittsburgh, 1981; repr. Ann Arbor, 1987).

Buysschaert, Anne-Caroline, "Roelant Savery als Tiermaler," in *Roelant Savery in seiner Zeit (1576-1639)*, exhibition catalogue (Cologne and Utrecht: Wallraf-Richartz-Museum, 1985), 51-4.

Callicott, J. Baird, "My Reply," in *Land, Value, Community: Callicott and Environmental Philosophy*, ed. by Wayne Ouderkirk and Jim Hill (Albany: State University of New York Press, 2002), 291-329.

Cassirer, Ernst, *The Individual and the Cosmos in Renaissance Philosophy*, trans. by Mario Domandi (New York: Harper and Row, 1964).

—, *The Philosophy of the Enlightenment*, trans. by Fritz C. A. Koelln and James P. Pettegrove (Boston: Beacon Press, 1960).

Charlton, D. G., *New Images of the Natural in France: A Study in European Cultural History 1750-1800* (Cambridge: Cambridge University Press, 1984).

Chinard, Gilbert, "Eighteenth Century Theories on America as a Human Habitat," *Proceedings of the American Philosophical Society*, 91 (1947), 27-57.

Chong, Alan, "'In 't verbeelden van Slachtdieren': Associaties en Betekenissen, Verbonden aan het Hollandse Veestuk in de Zeventiende Eeuw," in *Meesterlijk Vee: Nederlandse Veeschilders 1600-1900*, exhibition catalogue, ed. by C. Boschma (Zwolle: Waanders, and The Hague: Rijksdienst Beeldende Kunst, 1988), 56-86.

Clark, Kenneth, *Animals and Men: Their Relationship as Reflected in Western Art from Prehistory to the Present Day* (London: Thames and Hudson, 1977).

Clark, Robert T., Jr., *Herder: His Life and Thought* (Berkeley and Los Angeles: University of California Press, 1969).

Clark, Stephen R. L., "Hume, Animals and the Objectivity of Morals," *Philosophical Quarterly*, 35 (1985), 117-33.

Clarke, T. H., *The Rhinoceros from Dürer to Stubbs, 1515-1799* (London: Sotheby's, 1986).

Clutton-Brock, Juliet, *A Natural History of Domesticated Mammals* (Cambridge: Cambridge University Press, 1987).

Cohen, Esther, "Law, Folklore and Animal Lore," *Past and Present*, 110 (1986), 6-37.

Cohen, Sarah R., "Chardin's Fur: Painting, Materialism, and the Question of Animal Soul," *Eighteenth-Century Studies*, 38 (2004), 39-61.

—, "Life and Death in the Northern European Game Piece," in *Early Modern Zoology: The Construction of Animals in Science, Literature and the Visual Arts*, ed. by Karl A. E. Enenkel and Paul J. Smith, 2 vols. (Leiden and Boston: Brill, 2007), 2: 603-39.

—, "Thomas Gainsborough's Sensible Animals," in *Animals and Humans: Sensibility and Representation, 1650-1820*, ed. by Katherine M. Quinsey (Oxford: Voltaire Foundation, 2017), 191-218.

Cole, F. J., *A History of Comparative Anatomy: From Aristotle to the Eighteenth Century* (London: Macmillan, 1949), 270-305.

Cole, Lucinda, *Imperfect Creatures: Vermin, Literature, and the Sciences of Life, 1600-1740* (Ann Arbor: University of Michigan Press, 2016).

—, ed., "Animal, All Too Animal," Special issue of *The Eighteenth Century: Theory and Interpretation*, 52 (2011), 1-106.
Commager, Henry Steele, *The Empire of Reason: How Europe Imagined and America Realized the Enlightenment* (Garden City, NY: Anchor Press/Doubleday, 1977).
Cottingham, John, "A Brute to the Brutes? Descartes' Treatment of Animals," in *René Descartes, Critical Assessments*, vol. 4, ed. by G. J. D. Moyal (London and New York: Routledge, 1991), 323-31.
Craddock, Patricia B., *Young Edward Gibbon: Gentleman of Letters* (Baltimore and London: Johns Hopkins University Press, 1982).
Craig, Kenneth M., "Pieter Aertsen and 'The Meat Stall'," *Oud Holland*, 96 (1982), 1-15.
—, "Rembrandt and 'The Slaughtered Ox'," *Journal of the Warburg and Courtauld Institutes*, 46 (1983), 235-9.
Crosby, Alfred W., *Ecological Imperialism: The Biological Expansion of Europe, 900-1900* (Cambridge: Cambridge University Press, 1986).
Cummins, John, *The Hound and the Hawk: The Art of Medieval Hunting* (London: Weidenfeld and Nicolson, 1988).
Cuneo, Pia F., "(Un)stable Identities: Hippology and the Professionalization of Scholarship and Horsemanship in Early Modern Germany," in *Early Modern Zoology: The Construction of Animals in Science, Literature and the Visual Arts*, ed. by Karl A. E. Enenkel and Paul J. Smith, 2 vols. (Leiden and Boston: Brill, 2007), 2: 339-59.

Darnton, Robert, *The Great Cat Massacre and Other Episodes in French Cultural History* (Harmondsworth: Penguin Books, 1991).
Davis, Janet M., *The Gospel of Kindness: Animal Welfare and the Making of Modern America* (Oxford and New York: Oxford University Press, 2016).
Davis, Natalie Zemon, *Women on the Margins: Three Seventeenth-Century Lives* (Cambridge, MA and London: Harvard University Press, 1995).
Delort, Robert, *Les animaux ont une histoire* (Paris: Éditions du Seuil, 1984).
Dent, Anthony, *Animals in Art* (Oxford: Phaidon, 1976).
Dick, Steven J., *Plurality of Worlds: The Origins of the Extraterrestrial Life Debate from Democritus to Kant* (Cambridge: Cambridge University Press, 1982).
Dickenson, Victoria, "Meticulous Depiction: Animals in Art, 1400-1600," in *A Cultural History of Animals in the Renaissance*, ed. by Bruce Boehrer (Oxford and New York: Berg, 2007), 165-99.
Dijksterhuis, E. J., *The Mechanization of the World Picture: Pythagoras to Newton*, trans. by C. Dikshoorn (Princeton: Princeton University Press, 1986).
Dinzelbacher, Peter, "Animal Trials: A Multidisciplinary Approach," *Journal of Interdisciplinary History*, 32 (2002), 405-21.
Dobell, Clifford, *Antony van Leeuwenhoek and His 'Little Animals'* (London: Staples Press, 1932).
Donald, Diana, *Picturing Animals in Britain, 1750-1850* (New Haven and London: Yale University Press, 2007).
Duchet, Michèle, *Anthropologie et Histoire au siècle des lumières: Buffon, Voltaire, Rousseau, Helvétius, Diderot* (Paris: François Maspero, 1971), 235-49.
Dugas, René, "Sur le cartésianisme de Huygens," *Revue d'histoire des sciences*, 7 (1954), 22-33.
Dumas, Charles, *In het Zadel: Het Nederlands Ruiterportret van 1550 tot 1900*, exhibition catalogue (Leeuwarden: Fries Museum, 1979-80).
Duparc, F. J., "Philips Wouwerman, 1619-1668," *Oud Holland*, 107 (1993), 257-86.
Dvořák, Max, "Ein Stilleben des Bueckelaer oder Betrachtungen über die Entstehung der neuzeitigen Kabinettmalerei," *Jahrbuch der kunsthistorischen Sammlungen in Wien*, 36 (1923), 1-14.

Edelstein, Dan, *The Enlightenment: A Genealogy* (Chicago and London: University of Chicago Press, 2010).
Edwards, Peter, "Nature Bridled: The Treatment and Training of Horses in Early Modern England," in *Beastly Natures: Animals, Humans, and the Study of History*, ed. by Dorothee Brantz (Charlottesville and London: University of Virginia Press, 2010), 155-75.
Eisler, Colin, *Dürer's Animals* (Washington and London: Smithsonian Institution Press, 1991).
Elias, Norbert, *The Civilizing Process*, trans. by Edmund Jephcott (Oxford and Cambridge, MA: Blackwell, 1994).
Elzinga, Aant, "Christiaan Huygens' Theory of Research," in *Seventeenth-Century Natural Scientists*, ed. by Vere Chappell, *Essays on Early Modern Philosophers from Descartes and Hobbes to Newton and Leibniz*, 12 vols. (New York and London: Garland, 1992), 7: 147-66.
—, *On a Research Program in Early Modern Physics* (Göteborg: Läromedels-förlaget, 1972).
Emerson, Roger L., "Conjectural History and Scottish Philosophers," *Historical Papers / Communications Historiques*, 19 (1984), 63-90.
—, "The Scottish Enlightenment and the End of the Philosophical Society of Edinburgh," *British Journal for the History of Science*, 21 (1988), 33-66.
Evans, E. P., *The Criminal Prosecution and Capital Punishment of Animals* (London: Faber and Faber, 1987 [1906]).
Evensky, Jerry, *Adam Smith's Moral Philosophy: A Historical and Contemporary Perspective on Markets, Law, Ethics, and Culture* (Cambridge: Cambridge University Press, 2005).

Farbridge, Maurice H., *Studies in Biblical and Semitic Symbolism* (New York: Ktav, 1970).
Finkelstein, Andrea, *Harmony and the Balance: An Intellectual History of Seventeenth-Century English Economic Thought* (Ann Arbor: University of Michigan Press, 2000).
Fisher, Carl, "Politics and Porcine Representation: Multitudinous Swine in the British Eighteenth Century," *LIT: Literature Interpretation Theory*, 10 (1999), 303-26.
Floridi, Luciano, "Scepticism and Animal Rationality: The Fortune of Chrysippus' Dog in the History of Western Thought," *Archiv für Geschichte der Philosophie*, 79 (1997), 27-57.
Foucart-Walter, Élisabeth, and Pierre Rosenberg, *The Painted Cat: The Cat in Western Painting from the Fifteenth to the Twentieth Century* (New York: Rizzoli, 1988).
Freedberg, David, *Dutch Landscape Prints of the Seventeenth Century* (London: British Museum, 1980).
—, "The Hidden God: Image and Interdiction in the Netherlands in the Sixteenth Century," *Art History*, 5 (1982), 133-53.
—, "Science, Commerce, and Art: Neglected Topics at the Junction of History and Art History," in *Art in History, History in Art: Studies in Seventeenth-Century Dutch Culture*, ed. by David Freedberg and Jan de Vries, (Santa Monica: Getty Center for the History of Art and the Humanities, 1991), 376-428.
Freedberg, S. J., *Painting in Italy 1500-1600* (New Haven and London: Yale University Press, 1993).
French, Roger, *Dissection and Vivisection in the European Renaissance* (Aldershot: Ashgate, 1999).
Friedland, Paul, "Friends for Dinner: The Early Modern Roots of Modern Carnivorous Sensibilities," *History of the Present*, 1 (2011), 84-112.
Friedmann, Herbert, *The Symbolic Goldfinch: Its History and Significance in European Devotional Art* (Washington, D.C.: Pantheon Books, 1946).
Frierson, Patrick, "Metastandards in the Ethics of Adam Smith and Aldo Leopold," *Environmental Ethics*, 29 (2007), 171-91.
Frierson, Patrick R., "Adam Smith and the Possibility of Sympathy with Nature," *Pacific Philosophical Quarterly*, 87 (2006), 442-80.

Fromentin, Eugène, *The Masters of Past Time: Dutch and Flemish Painting from van Eyck to Rembrandt*, trans. by Anon. (Oxford: Phaidon Press, 1981 [1876]).
Fudge, Erica, *Brutal Reasoning: Animals, Rationality, and Humanity in Early Modern England* (Ithaca, NY: Cornell University Press, 2006).
—, "A Left-Handed Blow: Writing the History of Animals," in *Representing Animals*, ed. by Nigel Rothfels (Bloomington: Indiana University Press, 2002), 3-18.
—, *Perceiving Animals: Humans and Beasts in Early Modern English Culture* (Basingstoke and London: Macmillan Press, 2000).
—, "Two Ethics: Killing Animals in the Past and the Present," in The Animal Studies Group, *Killing Animals* (Urbana and Chicago: University of Illinois Press, 2006), 99-119.
—, Ed., *Renaissance Beasts: Of Animals, Humans, and Other Wonderful Creatures* (Urbana and Chicago: University of Illinois Press, 2004).

Garrett, Aaron, "Anthropology: the 'Original' of Human Nature," in *The Cambridge Companion to the Scottish Enlightenment*, ed. by Alexander Broadie (Cambridge: Cambridge University Press, 2003), 79-93.
—, "Francis Hutcheson and the Origin of Animal Rights," *Journal of the History of Philosophy*, 45 (2007), 243-65.
—, "Introduction," in Georges Louis Leclerc, Comte de Buffon, *Natural History, General and Particular*, trans. by William Smellie, 9 vols., 3rd ed. (London: A. Strahan and T. Cadell, 1791; repr. Bristol: Thoemmes Press, 2000), 1: v-xx.
Gasten, Andrea, "Dutch Still-Life Painting: Judgments and Appreciation," in *Still-Life in the Age of Rembrandt*, exhibition catalogue (Auckland: Auckland City Art Gallery, 1982), 13-25.
Gaudier, Jean-Pierre, and Jean-Jacques Heirwegh, "Jean-Paul Marana, *L'Espion du Grand Seigneur* et l'histoire des idées", *Études sur le xviiie siècle*, 8 (1981), 25-52.
Gay, Peter, *The Enlightenment: An Interpretation*, 2 vols. (New York and London: W. W. Norton, 1966-69).
Gelder, H. E. van, "Twee Braziliaanse Schildpadden door Albert Eckhout," *Oud Holland*, 75 (1960), 5-30.
Gelder, J. G. van, "The Animal and his 'Lettres de Noblesse'," *Apollo*, 104 (1976), 346-53.
George, Wilma, "Alive or Dead: Zoological Collections in the Seventeenth Century," in *The Origins of Museums*, ed. by Oliver Impey and Arthur MacGregor (Oxford: Clarendon Press, 1987), 179-87.
Gerbi, Antonello, *The Dispute of the New World: The History of a Polemic, 1750-1900*, trans. by Jeremy Moyle (Pittsburgh: University of Pittsburgh Press, 1973).
Gerson, H., and E.H. ter Kuile, *Art and Architecture in Belgium 1600 to 1800* (Harmondsworth: Penguin Books, 1960).
Geyl, Pieter, *The Netherlands in the Seventeenth Century*, 2 vols. (London and New York: E. Benn, 1961-64).
Gill, James E., "Theriophily in Antiquity: A Supplementary Account," *Journal of the History of Ideas*, 30 (1969), 401-12.
Glacken, Clarence J., *Traces on the Rhodian Shore: Nature and Culture in Western Thought from Ancient Times to the End of the Eighteenth Century* (Berkeley: University of California Press, 1967).
Goodwin, Craufurd D., "Ecologist Meets Economics: Aldo Leopold, 1887-1948," *Journal of the History of Economic Thought*, 30 (2008), 429-52.
Graham, Elspeth, "Reading, Writing, and Riding Horses in Early Modern England: James Shirley's *Hyde Park* (1632) and Gervase Markham's *Cavelarice* (1607)," in *Renaissance Beasts: Of Animals, Humans, and Other Wonderful Creatures*, ed. by Erica Fudge (Urbana and Chicago: University of Illinois Press, 2004), 116-37.

Griffin, Carl J., "Animal Maiming, Intimacy and the Politics of Shared Life: The Bestial and the Beastly in Eighteenth- and Early Nineteenth-Century England," *Transactions of the Institute of British Geographers*, 37 (2012), 301-16.

Griswold, Charles L., *Adam Smith and the Virtues of Enlightenment* (Cambridge: Cambridge University Press, 1999).

Grosjean, Ardis, "Toward an Interpretation of Pieter Aertsen's Profane Iconography," *Konsthistorisk Tidskrift*, 43 (1974), 121-43.

Grosser, Maurice, *The Painter's Eye* (New York: New American Library, 1956).

Guerrini, Anita, *The Courtiers' Anatomists: Animals and Humans in Louis XIV's Paris* (Chicago and London: University of Chicago Press, 2015).

—, "The Ethics of Animal Experimentation in Seventeenth-Century England," *Journal of the History of Ideas*, 50 (1989), 391-407.

—, *Experimenting with Humans and Animals: From Galen to Animal Rights* (Baltimore: Johns Hopkins University Press, 2003).

—, ed., with Domenico Bertoloni Meli, "Issue on Pre-Modern Animal Experimentation," *Journal of the History of Biology*, 46 (2013), 167-313.

Guichet, Jean-Luc, "Animality and Anthropology in Jean-Jacques Rousseau," trans. by Richard Byrne, in *A Cultural History of Animals in the Age of Enlightenment*, ed. by Matthew Senior (Oxford and New York: Berg, 2007), 145-56.

Gustafsson, Bo, "Nature and Economy," in *Nature and Society in Historical Context*, ed. by Mikuláš Teich, Roy Porter, and Bo Gustafsson (Cambridge: Cambridge University Press, 1997), 347-63.

Haak, Bob, *The Golden Age: Dutch Painters of the Seventeenth Century*, trans. E. Willems-Treeman (New York: Abrams, 1984).

Haakonssen, Lisbeth, *Medicine and Morals in the Enlightenment: John Gregory, Thomas Percival and Benjamin Rush* (Amsterdam and Atlanta: Rodopi, 1997).

Haraway, Donna, *The Companion Species Manifesto: Dogs, People, and Significant Otherness* (Chicago: Prickly Paradigm Press, 2003).

Hargraves, Neil, "Beyond the Savage Character: Mexicans, Peruvians, and the 'Imperfectly Civilized' in William Robertson's 'History of America'," in *The Anthropology of the Enlightenment*, ed. by Larry Wolff and Marco Cipolloni (Stanford: Stanford University Press, 2007), 103-18.

Harman, P. M., *The Culture of Nature in Britain 1680-1860* (New Haven and London: Yale University Press, 2009).

Harrison, Peter, "Descartes on Animals," *Philosophical Quarterly*, 42 (1992), 219-27.

—, *The Fall of Man and the Foundations of Science* (Cambridge: Cambridge University Press, 2007).

—, "The Virtues of Animals in Seventeenth-Century Thought," *Journal of the History of Ideas*, 59 (1998), 463-84.

Harwood, Dix, *Love for Animals and How it Developed in Great Britain* (New York: Columbia University, 1928).

Havens, George R., "Voltaire, Rousseau, and the 'Lettre sur la Providence,'" *PMLA*, 59 (1944), 109-30.

Hastings, Hester, *Man and Beast in French Thought of the Eighteenth Century* (Baltimore: Johns Hopkins Press, 1936).

Heckscher, Eli F., *Mercantilism*, trans. by Mendel Shapiro, ed. by E. F. Söderlund, 2 vols. (London: George Allen & Unwin, and New York: Macmillan, 1962).

Herlitz, Lars, "Art and Nature in Pre-Classical Economics of the Seventeenth and Eighteenth Centuries," in *Nature and Society in Historical Context*, ed. by Mikuláš Teich, Roy Porter, and Bo Gustafsson (Cambridge: Cambridge University Press, 1997), 163-75.

Heugten, Sjraar van, "Grazing Models: Aspects of the Dutch Cattle Piece," trans. by Ruth Koenig, in *Meesterlijk Vee: Nederlandse Veeschilders 1600-1900*, exhibition catalogue, ed. by C. Boschma (Zwolle: Waanders, and The Hague: Rijksdienst Beeldende Kunst, 1988), 266-77.
Hont, Istvan, *Jealousy of Trade: International Competition and the Nation-State in Historical Perspective* (Cambridge, MA, and London: Harvard University Press, 2005).
—, "The Language of Sociability and Commerce: Samuel Pufendorf and the Theoretical Foundations of the 'Four-Stages Theory'," in *The Languages of Political Theory in Early-Modern Europe*, ed. by Anthony Pagden (Cambridge: Cambridge University Press, 1987), 253-76.
Höpfl, H. M., "From Savage to Scotsman: Conjectural History in the Scottish Enlightenment," *Journal of British Studies*, 17 (1978), 19-40.
Hoquet, Thierry, *Buffon: histoire naturelle et philosophie* (Paris: Honoré Champion Éditeur, 2005).
Howe, William Norton, *Animal Life in Italian Painting* (London: G. Allen, 1912).
Huxley, Thomas H., "On the Hypothesis that Animals are Automata and its History," in *Collected Essays (1893-1894)*, Vol. I: *Method and Results* (Hildesheim and New York: G. Olms, 1970), 199-250.

Israel, Jonathan, *Democratic Enlightenment: Philosophy, Revolution, and Human Rights 1750-1790* (Oxford: Oxford University Press, 2011).
—, *Enlightenment Contested: Philosophy, Modernity, and the Emancipation of Man 1670-1752* (Oxford: Oxford University Press, 2006).
—, *Radical Enlightenment: Philosophy and the Making of Modernity 1650-1750* (Oxford: Oxford University Press, 2001).
—, *A Revolution of the Mind: Radical Enlightenment and the Intellectual Origins of Modern Democracy* (Princeton and Oxford: Princeton University Press, 2010).
—, *Revolutionary Ideas: An Intellectual History of the French Revolution from* The Rights of Man *to Robespierre* (Oxford and Princeton: Princeton University Press, 2014).

Jackson, Christine E., *Great Bird Paintings of the World*, Vol. 1: *The Old Masters* (Woodbridge: Antique Collectors' Club, 1993).
Jansen, Guido M. C., with contributions by Josefine Leistra and Kees C. J. Stal, *Connoisseurship – Bredius | Jan Steen and the Mauritshuis*, exhibition catalogue, Museum Bredius, The Hague (The Hague: Bredius Society, 2014).
Janson, H. W., *Apes and Ape Lore in the Middle Ages and the Renaissance* (London: Warburg Institute, 1952).
Jarcho, Saul, "Giuseppe Zambeccari: A Seventeenth-Century Pioneer in Experimental Physiology and Surgery," *Bulletin of the History of Medicine*, 9 (1941), 144-67.
Jenner, Mark S. R., "The Great Dog Massacre," in *Fear in Early Modern Society*, ed. by William G. Naphy and Penny Roberts (Manchester and New York: Manchester University Press, 1997), 44-61.
Jongh, Eddy de, "The Iconological Approach to Seventeenth-Century Dutch Painting," in *The Golden Age of Dutch Painting in Historical Perspective*, ed. by Frans Grijzenhout and Henk van Veen, trans. by Andrew McCormick (Cambridge: Cambridge University Press, 1999), 200-23.
—, "The Interpretation of Still-Life Paintings: Possibilities and Limits," in *Still-Life in the Age of Rembrandt*, exhibition catalogue (Auckland: Auckland City Art Gallery, 1982), 27-37.
—, "Realism and Seeming Realism in Seventeenth-Century Dutch Painting," in *Looking at Seventeenth-Century Dutch Art: Realism Reconsidered*, ed. by Wayne Franits (Cambridge: Cambridge University Press, 1997), 21-56.
—, "Review of Svetlana Alpers, *The Art of Describing*," *Simiolus*, 14 (1984), 51-9.

—, "Some Notes on Interpretation," in *Art in History, History in Art: Studies in Seventeenth-Century Dutch Culture*, ed. by David Freedberg and Jan de Vries, (Santa Monica: Getty Center for the History of Art and the Humanities, 1991), 118-36.

Jonsson, Fredrik Albritton, *Enlightenment's Frontier: The Scottish Highlands and the Origins of Environmentalism* (New Haven and London: Yale University Press, 2013).

—, "Natural History and Improvement: The Case of Tobacco," in *Mercantilism Reimagined: Political Economy in Early Modern Britain and Its Empire*, ed. by Philip J. Stern and Carl Wennerlind (Oxford: Oxford University Press, 2014), 117-33.

Joppien, R., "The Dutch Vision of Brazil, Johan Maurits and his Artists," in *Johan Maurits van Nassau-Siegen, 1604-1679: A Humanist Prince in Europe and Brazil*, ed. by E. van den Boogaart, H. R. Hoetink, and P. J. P. Whitehead (The Hague: Johan Maurits van Nassau Stichting, 1979), 297-376.

Kafker, Frank A., "William Smellie's Edition of the *Encyclopaedia Britannica*," in *Notable Encyclopedias of the Late Eighteenth Century: Eleven Successors of the Encyclopédie*, ed. by Frank A. Kafker (Oxford: The Voltaire Foundation, 1994), 145-82.

Kalof, Linda, *Looking at Animals in Human History* (London: Reaktion Books, 2007).

Kean, Hilda, *Animal Rights: Political and Social Change in Britain since 1800* (London: Reaktion Books, 1998).

Kelly, Christopher, "Rousseau's 'Peut-Etre': Reflections on the Status of the State of Nature," *Modern Intellectual History*, 3 (2006), 75-83.

Kemp, Martin, *The Human Animal in Western Art and Science* (Chicago and London: University of Chicago Press, 2007).

Kenyon-Jones, Christine, *Kindred Brutes: Animals in Romantic Period Writing* (Aldershot: Ashgate, 2001).

Kerestman, Katherine, "Breaking the Shackles of the Great Chain of Being and Liberating Compassion in the Eighteenth Century," *1650-1850: Ideas, Aesthetics, and Inquiries in the Early Modern Era*, 3 (1997), 57-76.

Kete, Kathleen, *The Beast in the Boudoir: Petkeeping in Nineteenth-Century Paris* (Berkeley: University of California Press, 1994).

—, Ed., *A Cultural History of Animals in the Age of Empire* (Oxford and New York: Berg, 2007).

Kettering, Alison M., *The Dutch Arcadia: Pastoral Art and Its Audience in the Golden Age* (Totowa, NJ: Allanheld and Schram, 1983).

Keynes, Geoffrey, *The Library of Edward Gibbon: A Catalogue of His Books* (London: Jonathan Cape, 1940).

Klemme, Heiner F., "[Review of books on John Gregory by Laurence B. McCullough and Lisbeth Haakonssen]," *Journal of the History of Philosophy*, 37 (1999), 535-8.

Klingender, Francis, *Animals in Art and Thought to the End of the Middle Ages* (London: Routledge and Kegan Paul, 1971).

Knight, David, "Celestial Worlds Discover'd," *Durham University Journal*, 58, n.s. 27 (1965), 23-9.

Knight, David M., "Uniformity and Diversity of Nature in 17th Century Treatises on Plurality of Worlds," *Organon*, 4 (1967), 61-8.

Koerner, Joseph Leo, *The Moment of Self-Portraiture in German Renaissance Art* (Chicago and London: University of Chicago Press, 1993), 426-37.

Koreny, Fritz, *Albrecht Dürer and the Animal and Plant Studies of the Renaissance*, trans. by P. Marwood and Y. Shapiro (Boston: Little, Brown and Co., 1988).

Kors, Alan Charles, *D'Holbach's Coterie: An Enlightenment in Paris* (Princeton: Princeton University Press, 1976).

Koslow, Susan, *Frans Snyders* (Antwerp: Mercator Fonds, 1995).
Koyré, Alexandre, *From the Closed World to the Infinite Universe* (Baltimore: Johns Hopkins University Press, 1994).
Kruyfhooft, Cécile, and Simone Buys, "P. P. Rubens et la Peinture Animaliere," *Zoo Antwerpen*, supplement (1977).
Kula, E., *History of Environmental Economic Thought* (London and New York: Routledge, 1998).
Kulstad, Mark, "Leibniz, Animals, and Apperception," *Studia Leibnitiana*, 13 (1981), 25-60.

LaFreniere, Gilbert F., *The Decline of Nature: Environmental History and the Western Worldview* (Palo Alto: Academica Press, 2008).
Landry, Donna, *The Invention of the Countryside: Hunting, Walking and Ecology in English Literature, 1671-1831* (Basingstoke: Palgrave, 2001).
Larsen, Erik, "The Duality of Flemish and Dutch Art in the Seventeenth Century," in *Stil und Überlieferung in der Kunst des Abendlandes, Akten des 21. Internationalen Kongresses für Kunstgeschichte in Bonn 1964, Band III: Theorien und Probleme* (Berlin: Gebr. Mann, 1967), 59-65.
Leach, Edmund, "Anthropological Aspects of Language: Animal Categories and Verbal Abus," in *New Directions in the Study of Language*, ed. by Eric H. Lenneberg (Cambridge, MA: MIT Press, 1964), 23-63.
Leigh, R. A., "From the *Inégalité* to *Candide*: Notes on a Desultory Dialogue between Rousseau and Voltaire (1755-1759)," in *The Age of the Enlightenment: Studies Presented to Theodore Besterman*, ed. by W. H. Barber, J. H. Brumfitt, R. A. Leigh, R. Shackleton, and S. S. B. Taylor (Edinburgh: Oliver and Boyd, 1967), 66-92.
—, "Rousseau's Letter to Voltaire on Optimism," *Studies on Voltaire and the Eighteenth Century*, 30 (1964), 247-309.
Leshem, Dotan, "Oikonomia Redefined," *Journal of the History of Economic Thought*, 35 (2013), 43-61.
Lévi-Strauss, Claude, *The Savage Mind* (Chicago: University of Chicago Press, 1969).
Lewine, Carol F., *The Sistine Chapel Walls and the Roman Liturgy* (University Park: Pennsylvania State University Press, 1993).
Lewis, Rhodri, "Thinking with Animals in the Early Royal Society: The Case of Sir William Petty," in *Ethical Perspectives on Animals in the Renaissance and Early Modern Period*, Micrologus' Library 55, ed. by Cecilia Muratori and Burkhard Dohm (Florence: Sismel-Edizioni del Galluzzo, 2013), 231-56.
Lieberman, David, "Adam Smith on Justice, Rights, and Law," in *The Cambridge Companion to Adam Smith*, ed. by Knud Haakonssen (Cambridge: Cambridge University Press, 2006), 214-45.
Liedtke, Walter, *The Royal Horse and Rider: Painting, Sculpture, and Horsemanship 1500-1800* (New York: Abaris Books, 1989).
Lindeboom, G. A., "Dog and Frog: Physiological Experiments at Leiden During the Seventeenth Century," in *Leiden University in the Seventeenth Century: An Exchange of Learning*, ed. by Th. H. Lunsingh Scheurleer and G. H. M. P. Meyjes (Leiden: University of Leiden Press / Brill, 1975), 279-93.
Lloyd, Joan B., *African Animals in Renaissance Literature and Art* (Oxford: Clarendon Press, 1971).
Lodi, Silvano, *Natura Morta Italiana: Italian Still Life Painting from Four Centuries, The Silvano Lodi Collection*, exhibition catalogue (Jerusalem: Israel Museum, 1994).
Loisel, Gustave, *Histoire des ménageries de l'antiquité à nos jours*, 3 vols. (Paris: Doin, 1912).
Lovejoy, Arthur O., *The Great Chain of Being: A Study of the History of an Idea* (New York: Harper and Row, 1960 [1936]).
—, "The Supposed Primitivism of Rousseau's *Discourse on Inequality*," in *Essays in the History of Ideas* (Baltimore: Johns Hopkins Press, 1961), 14-37.

Lovejoy, Arthur O., and George Boas, *Primitivism and Related Ideas in Antiquity* (Baltimore and London: Johns Hopkins University Press, 1997 [1935]).

Luyendijk-Elshout, A. M., "Introduction," in Frederik Ruysch, *Dilucidatio Valvularum in Vasis Lymphaticis et Lacteis*, trans. by I. Seeger-Wolf (Nieuwkoop: De Graaf, 1964), 7-44.

MacDonogh, Katharine, *Reigning Cats and Dogs* (New York: St. Martin's Press, 1999).

MacInnes, Ian, "Mastiffs and Spaniels: Gender and Nation in the English Dog," *Textual Practice*, 17 (2003), 21-40.

MacIntosh, J. J., "Animals, Morality and Robert Boyle," *Dialogue*, 35 (1996), 435-72.

Maehle, Andreas-Holger, "Cruelty and Kindness to the 'Brute Creation', Stability and Change in the Ethics of the Man-Animal Relationship, 1600-1850," in *Animals and Human Society: Changing Perspectives*, ed. by Aubrey Manning and James Serpell (London: Routledge, 1994), 81-105.

Maehle, Andreas-Holger, and Ulrich Tröhler, "Animal Experimentation from Antiquity to the End of the Eighteenth Century: Attitudes and Arguments," in *Vivisection in Historical Perspective*, ed. by Nicolaas A. Rupke (London: Routledge, 1987), 14-47.

Malcolmson, Robert, and Stephanos Mastoris, *The English Pig: A History* (London and Rio Grande: Hambledon Press, 1998).

Maluf, N. S. R., "History of Blood Transfusion," *Journal of the History of Medicine and Allied Sciences*, 9 (1954), 59-107.

Mancilla, Alejandra, "Nonhuman Animals in Adam Smith's Moral Theory," *Between the Species*, 9 (August 2009), accessed in May 2016 at: http://digitalcommons.calpoly.edu/bts/vol13/iss9/2/

Mangum, Teresa, "Dog Years, Human Fears," in *Representing Animals*, ed. by Nigel Rothfels (Bloomington: Indiana University Press, 2002), 35-47.

Mani, Nikolaus, "Jean Riolan II (1580-1657) and Medical Research," *Bulletin of the History of Medicine*, 42 (1974), 121-44.

Marouby, Christian, "Adam Smith and the Anthropology of the Enlightenment: The 'Ethnographic' Sources of Economic Progress," in *The Anthropology of the Enlightenment*, ed. by Larry Wolff and Marco Cipolloni (Stanford: Stanford University Press, 2007), 85-102.

Marx, Leo, *The Machine in the Garden: Technology and the Pastoral Ideal in America* (New York: Oxford University Press, 1964).

McBurney, William H., "The Authorship of *The Turkish Spy*," *PMLA*, 72 (1957), 915-35.

McColley, Grant, "The Seventeenth-Century Doctrine of a Plurality of Worlds," *Annals of Science*, 1 (1936), 385-430.

McCormick, Ted, *William Petty and the Ambitions of Political Arithmetic* (Oxford: Oxford University Press, 2009).

McCullough, Laurence B., "Hume's Influence on John Gregory and the History of Medical Ethics," *Journal of Medicine and Philosophy*, 24 (1999), 376-95.

—, *John Gregory and the Invention of Professional Medical Ethics and the Profession of Medicine* (Dordrecht: Kluwer Academic Publishers, 1998).

—, "John Gregory's Medical Ethics and Humean Sympathy," in *The Codification of Medical Morality: Historical and Philosophical Studies of the Formalization of Western Medical Morality in the Eighteenth and Nineteenth Centuries*, Vol. 1: *Medical Ethics and Etiquette in the Eighteenth Century*, ed. by Robert Baker, Dorothy Porter, and Roy Porter (Dordrecht: Kluwer Academic Publishers, 1993), 145-60.

McMahon, Darrin. M., "What are Enlightenments?" *Modern Intellectual History*, 4 (2007), 601-16.

McShane, Clay, and Joel A. Tarr, "The Horse in the Nineteenth-Century American City," in *Beastly Natures: Animals, Humans, and the Study of History*, ed. by Dorothee Brantz (Charlottesville and London: University of Virginia Press, 2010), 227-45.

Meek, Ronald L., *Social Science and the Ignoble Savage* (Cambridge: Cambridge University Press, 1976).
Melehy, Hassan, "Silencing the Animals: Montaigne, Descartes, and the Hyperbole of Reason," *symplokē*, 13 (2005), 263-82.
Meli, Domenico Bertoloni, and Anita Guerrini, eds., "The Representation of Animals in the Early Modern Period," *Annals of Science*, 67 (2010), 299-429.
Meyer, A. W., "Leeuwenhoek as Experimental Biologist," *Osiris*, 3 (1937), 103-22.
Miedema, Hessel, "The Appreciation of Paintings around 1600," in *Dawn of the Golden Age: Northern Netherlandish Art, 1580-1620*, exhibition catalogue, ed. by Ger Luijten, Ariane van Suchtelen, Reinier Baarsen, and Walter Kloek (Amsterdam: Rijksmusem, and Zwolle: Waanders, 1993), 122-35.
—, "Karel van Mander's Grondt der Edel Vry Schilder-Const ('Foundations of the Noble and Free Art of Painting')," *Journal of the History of Ideas*, 34 (1973), 653-68.
Miles, Murray, "Leibniz on Apperception and Animal Souls," *Dialogue*, 33 (1994), 701-24.
Miller, Michael R., "Descartes on Animals Revisited," *Journal of Philosophical Research*, 38 (2013), 89-114.
Mirowski, Philip, "The Realms of the Natural," in *Natural Images in Economic Thought: "Markets Read in Tooth and Claw"*, ed. by Philip Mirowski (Cambridge: Cambridge University Press, 1994), 451-83.
Mizelle, Brett, *Pig* (London: Reaktion Books, 2011).
Momigliano, Arnaldo, "Gibbon's Contribution to Historical Method," in *Studies in Historiography* (London: Weidenfeld and Nicolson, 1966), 40-55.
Montes, Leonidas, "*Das Adam Smith Problem*: Its Origins, the Stages of the Current Debate, and One Implication for Our Understanding of Sympathy," *Journal of the History of Economic Thought*, 25 (2003), 63-90.
Morford, Mark, *Stoics and Neostoics: Rubens and the Circle of Lipsius* (Princeton: Princeton University Press, 1991).
Motley, John Lothrop, *The Rise of the Dutch Republic*, 3 vols. (London: John Murray, 1903-1904).
Moxey, Keith P. F., "The 'Humanist' Market Scenes of Joachim Beuckelaer: Moralizing Exempla or 'Slices of Life'?" *Jaarboek van het Koninklijk Museum voor Schone Kunsten Antwerpen* (1976), 109-87.
—, *Pieter Aertsen, Joachim Beuckelaer, and the Rise of Secular Painting in the Context of the Reformation* (Ph.D. diss., University of Chicago, 1974; repr. New York and London: Garland, 1977).
Müllenmeister, Kurt J., *Meer und Land im Licht des 17. Jahrhunderts*, vols. 2 and 3: *Tierdarstellungen in Werken niederländischer Künstler* (Bremen: Carl Schünemann, 1978, 1981).
Munda, Giuseppe, "Environmental Economics, Ecological Economics, and the Concept of Sustainable Development," *Environmental Values*, 6 (1997), 213-33.
Munitz, Milton K., "One Universe or Many?" in *Roots of Scientific Thought: A Cultural Perspective*, ed. by Philip P. Wiener and Aaron Noland (New York: Basic Books, 1957), 593-617.
Muratori, Cecilia, and Burkhard Dohm, eds., *Ethical Perspectives on Animals in the Renaissance and Early Modern Period*, Micrologus' Library 55 (Florence: Sismel-Edizioni del Galluzzo, 2013).
Muthu, Sankar, *Enlightenment Against Empire* (Princeton and Oxford: Princeton University Press, 2003).

Nash, Roderick, *Wilderness and the American Mind* (New Haven and London: Yale University Press, 1973).
Nelson, Holly Faith, "'Worms in the Dull Earth of Ignorance': Zoosemiotics and Sexual Politics in the Works of Margaret Cavendish, Duchess of Newcastle," *English Language Notes*, 39 (2002), 12-24.
Newman, Lex, "Unmasking Descartes's Case for the Bête Machine Doctrine," *Canadian Journal of Philosophy*, 31 (2001), 389-425.

Newmyer, Stephen T., "Animals in Ancient Philosophy: Conceptions and Misconceptions," in *A Cultural History of Animals in Antiquity*, ed. by Linda Kalof (Oxford and New York: Berg, 2007), 151-74.

Nicklas, Pascal, "Letters of the Self: Otherness and Epistolary Writing in Giovanni Paolo Marana's *Letters Writ by a Turkish Spy*," in *Narrative Strategies in Early English Fiction*, ed. by Wolfgang Görtschacher and Holger Klein (Lewiston and Salzburg: Edwin Mellen Press, 1995), 353-65.

Nicolson, Marjorie, and G. S. Rousseau, *"This Long Disease, My Life": Alexander Pope and the Sciences* (Princeton: Princeton University Press, 1968), 93-109.

Norton, Marcy, "The Chicken or the *Iegue*: Human-Animal Relationships and the Columbian Exchange," *American Historical Review*, 120 (2015), 28-60.

Nuyen, A. T., "Hume on Animals and Morality," *Philosophical Papers*, 27 (1998), 93-106.

O'Brien, Karen, *Narratives of Enlightenment: Cosmopolitan History from Voltaire to Gibbon* (Cambridge: Cambridge University Press, 1997).

—, "The Return of the Enlightenment," *American Historical Review*, 115 (2010), 1426-35.

O'Connor, Martin, "John Stuart Mill's Utilitarianism and the Social Ethics of Sustainable Development," *European Journal of the History of Economic Thought*, 4 (1997), 478-506.

Old Master Paintings and Works of Art from the Bentinck-Thyssen Collection (London: Sotheby's catalogue, December 6, 1995).

Oz-Salzberger, Fania, "New Approaches Towards a History of the Enlightenment: Can Disparate Perspectives Make a General Picture?" *Tel Aviver Jahrbuch für Deutsche Geschichte*, 29 (2000), 171-82.

—, *Translating the Enlightenment: Scottish Civic Discourse in Eighteenth-Century Germany* (Oxford: Clarendon Press, 1995).

Pächt, Otto, "Early Italian Nature Studies and the Early Calendar Landscape," *Journal of the Warburg and Courtauld Institutes*, 13 (1950), 13-47.

Pagden, Anthony, *Lords of All the World: Ideologies of Empire in Spain, Britain and France c. 1500-c. 1800* (New Haven and London: Yale University Press, 1995).

Palmeri, Frank, "Conjectural History and the Origins of Sociology," *Studies in Eighteenth Century Culture*, 37 (2008), 1-21.

—, "A Profusion of Dead Animals: Autocritique in Seventeenth-Century Flemish Gamepieces," *Journal for Early Modern Cultural Studies*, 16 (2016), 50-77.

—, ed., *Humans and Other Animals in Eighteenth-Century British Culture: Representation, Hybridity, Ethics* (Aldershot and Burlington: Ashgate, 2006).

Panichi, Nicola, "Montaigne and Animal Ethics," in *Ethical Perspectives on Animals in the Renaissance and Early Modern Period*, ed. by Cecilia Muratori and Burkhard Dohm, Micrologus' Library 55 (Florence: Sismel-Edizioni del Galluzzo, 2013), 91-110.

Papy, Jan, "Lipsius and His Dogs: Humanist Tradition, Iconography and Rubens's *Four Philosophers*," *Journal of the Warburg and Courtauld Institutes*, 62 (1999), 167-98.

Parham, John, "What is (Ecological) 'Nature'? John Stuart Mill and the Victorian Perspective," in *Culture, Creativity and Environment: New Environmentalist Criticism*, ed. by Fiona Becket and Terry Gifford (Amsterdam and New York: Rodopi, 2007), 37-54.

Parr, Charles McKew, *Jan van Linschoten: The Dutch Marco Polo* (New York: T. Y. Crowell, 1964).

Partsch, Susanna, *Franz Marc 1880-1916*, trans. by Karen Williams (Cologne: Taschen, 1991).

Passmore, John, *Man's Responsibility for Nature: Ecological Problems and Western Traditions* (London: Duckworth, 1974).

Perkins, David, *Romanticism and Animal Rights* (Cambridge: Cambridge University Press, 2003).

Philbrick, Thomas, *St. John de Crèvecoeur* (New York: Twayne Publishers, 1970).
Phillips, Mark Salber, *Society and Sentiment: Genres of Historical Writing in Britain, 1740-1820* (Princeton: Princeton University Press, 2000).
Pitson, Tony, "Hume on Morals and Animals," *British Journal for the History of Philosophy*, 11 (2003), 639-55.
Pitts, Jennifer, *A Turn to Empire: The Rise of Imperial Liberalism in Britain and France* (Princeton: Princeton University Press, 2005), 25-58.
Pocock, J. G. A., *Barbarism and Religion*, Vol. 2: *Narratives of Civil Government* (Cambridge: Cambridge University Press, 1999).
—, *Barbarism and Religion*, Vol. 4: *Barbarians, Savages and Empires* (Cambridge: Cambridge University Press, 2005).
—, *Barbarism and Religion*, Vol. 5: *Religion: The First Triumph* (Cambridge: Cambridge University Press, 2010).
—, "Gibbon and the Shepherds: The Stages of Society in the *Decline and Fall*," *History of European Ideas*, 2 (1981), 193-202.
—, "Historiography and Enlightenment: A View of Their History," *Modern Intellectual History*, 5 (2008), 83-96.
Porter, Roy, "The Urban and the Rustic in Enlightenment London," in *Nature and Society in Historical Context*, ed. by Mikuláš Teich, Roy Porter, and Bo Gustafsson (Cambridge: Cambridge University Press, 1997), 176-94.
Porter, Roy, and Mikuláš Teich, eds., *The Enlightenment in National Context* (Cambridge: Cambridge University Press, 1981).
Preece, Rod, *Animals and Nature: Cultural Myths, Cultural Realities* (Vancouver: UBC Press, 1999).
—, "Darwinism, Christianity, and the Great Vivisection Debate," *Journal of the History of Ideas*, 64 (2003), 399-419.
Price, Matt, "Economics, Ecology, and the Value of Nature," in *The Moral Authority of Nature*, ed. by Lorraine Daston and Fernando Vidal (Chicago and London: University of Chicago Press, 2004), 182-204.

Quinsey, Katherine M., ed., *Animals and Humans: Sensibility and Representation, 1650-1820* (Oxford: Voltaire Foundation, 2017).

Raber, Karen, "From Sheep to Meat, From Pets to People: Animal Domestication 1600-1800," in *A Cultural History of Animals in the Age of Enlightenment*, ed. by Matthew Senior (Oxford and New York: Berg, 2007), 73-99.
Raber, Karen, and Treva J. Tucker, eds., *The Culture of the Horse: Status, Discipline, and Identity in the Early Modern World* (New York and Basingstoke: Palgrave Macmillan, 2005).
Rachels, James, *Created from Animals: The Moral Implications of Darwinism* (Oxford and New York: Oxford University Press, 1990).
Radner, Daisie, and Michael Radner, *Animal Consciousness* (Buffalo: Prometheus Books, 1996).
Raphael, D. D., *The Impartial Spectator: Adam Smith's Moral Philosophy* (Oxford: Clarendon Press, 2006).
Rausing, Lisbet, "Underwriting the Oeconomy: Linnaeus on Nature and Mind," in *Oeconomies in the Ages of Newton*, ed. by Margaret Schabas and Neil De Marchi, Annual Supplement to vol. 35, *History of Political Economy* (Durham and London: Duke University Press, 2003), 173-203.
Reill, Peter Hanns, "Buffon and Historical Thought in Germany and Great Britain," in *Buffon 88: Actes du Colloque international pour le bicentenaire de la mort de Buffon (Paris, Montbard, Dijon, 14-22 juin 1988)*, ed. by Michel Delsol (Paris: VRIN, 1992), 667-79.

—, *Vitalizing Nature in the Enlightenment* (Berkeley: University of California Press, 2005).
Resl, Brigitte, "Beyond the Ark: Animals in Medieval Art," in *A Cultural History of Animals in the Medieval Age*, ed. by Brigitte Resl (Oxford and New York: Berg, 2007), 179-201.
Reutersward, Patrik, "The Dog in the Humanist's Study," *Konsthistorisk Tidskrift*, 50 (1981), 53-69.
Ridley, Glynis, ed., "Animals in the Eighteenth Century," *Journal for Eighteenth-Century Studies*, 33 (2010), 427-683.
Riskin, Jessica, "The 'Spirit of System' and the Fortunes of Physiocracy," *History of Political Economy*, 35, Supplement (2003), 42-73.
Ritvo, Harriet, *The Animal Estate: The English and Other Creatures in the Victorian Age* (Cambridge, MA: Harvard University Press, 1987).
—, "Animal Planet," *Environmental History*, 9 (2004), 204-20.
Robertson, John, *The Case for the Enlightenment: Scotland and Naples 1680-1760* (Cambridge: Cambridge University Press, 2005).
Robbins, Louise E., *Elephant Slaves and Pampered Parrots: Exotic Animals in Eighteenth-Century Paris* (Baltimore and London: Johns Hopkins University Press, 2002).
Roger, Jacques, *Buffon: A Life in Natural History*, trans. by Sarah Lucille Bonnefoi, ed. by L. Pearce Williams (Ithaca and London: Cornell University Press, 1997).
Rosenberg, Jakob, Seymour Slive, and E.H. ter Kuile, *Dutch Art and Architecture 1600-1800* (New Haven and London: Yale University Press, 1993).
Rosenblum, Robert, *The Dog in Art from Rococo to Post-Modernism* (New York: Abrams, 1988).
Rosenfield, Leonora Cohen, *From Beast-Machine to Man-Machine: Animal Soul in French Letters from Descartes to La Mettrie* (New York: Octagon Books, 1968 [1940]).
Rossi, Paolo, "Nobility of Man and Plurality of Worlds," trans. by Arthur Brickmann, in *Science, Medicine and Society in the Renaissance: Essays to Honor Walter Pagel*, ed. by Allen G. Debus, 2 vols. (London: Heinemann, 1972), 2: 131-62.
Rothschild, Emma, *Economic Sentiments: Adam Smith, Condorcet, and the Enlightenment* (Cambridge, MA, and London: Harvard University Press, 2001).
—, "Maintaining (Environmental) Capital Intact," *Modern Intellectual History*, 8 (2011), 193-212.
Rupke, Nicolaas A., ed., *Vivisection in Historical Perspective* (London: Routledge, 1987).
Rupp, Jan C. C., "Matters of Life and Death: The Social and Cultural Conditions of the Rise of Anatomical Theatres, with Special Reference to Seventeenth-Century Holland," *History of Science*, 28 (1990), 263-87.

Sahlins, Peter, "The Beast Within: Animals in the First Xenotransfusion Experiments in France, ca. 1667-68," *Representations*, 129 (2015), 25-55.
—, "The Royal Menageries of Louis XIV and the Civilizing Process Revisited," *French Historical Studies*, 35 (2012), 237-67.
Salisbury, Joyce E., *The Beast Within: Animals in the Middle Ages* (New York and London: Routledge, 1994).
Schabas, Margaret, *The Natural Origins of Economics* (Chicago and London: University of Chicago Press, 2005).
Schaffer, Simon, "The Earth's Fertility as a Social Fact in Early Modern Britain," in *Nature and Society in Historical Context*, ed. by Mikuláš Teich, Roy Porter, and Bo Gustafsson (Cambridge: Cambridge University Press, 1997), 124-47.
Schama, Simon, *The Embarrassment of Riches: An Interpretation of Dutch Culture in the Golden Age* (Berkeley: University of California Press, 1988).
Scheurleer, Th. H. L., "Early Dutch Cabinets of Curiosities," in *The Origins of Museums*, ed. by Oliver Impey and Arthur MacGregor (Oxford: Clarendon Press, 1987), 115-20.

Schloss, Christine, "The Early Italianate Genre Paintings by Jan Weenix (ca. 1642-1719)," *Oud Holland*, 97 (1983), 69-97.
Schneider, Norbert, *The Art of the Still Life: Still Life Painting in the Early Modern Period*, trans. by Hugh Beyer (Cologne: Taschen, 1990).
Schouten, J., "Johannes Walaeus (1604-1649) and His Experiments on the Circulation of the Blood," *Journal of the History of Medicine and Allied Sciences*, 29 (1974), 259-79.
Seibert, Margaret Armbrust, "A Political and Pictorial Tradition Used in Gustave Courbet's *Real Allegory*," *Art Bulletin*, 65 (1983), 311-16.
Seidengart, Jean, "Les théories cosmologiques de Christiaan Huygens," in *Huygens et la France, table ronde du Centre National de la Recherche Scientifique, Paris 27-29 mars 1979* (Paris: Librairie philosophique J. Vrin, 1982), 209-22.
Seidler, Michael J., "Hume and the Animals," *Southern Journal of Philosophy*, 15 (1977), 361-72.
Senior, Matthew, "The Ménagerie and the Labyrinthe: Animals at Versailles, 1662-1792," in *Renaissance Beasts: Of Animals, Humans, and Other Wonderful Creatures*, ed. by Erica Fudge (Urbana and Chicago: University of Illinois Press, 2004), 208-32.
—, "The Souls of Men and Beasts, 1630-1764," in *A Cultural History of Animals in the Age of Enlightenment*, ed. by Matthew Senior (Oxford and New York: Berg, 2007), 23-45.
—, ed., *A Cultural History of Animals in the Age of Enlightenment* (Oxford and New York: Berg, 2007).
Shannon, Laurie, *The Accommodated Animal: Cosmopolity in Shakespearean Locales* (Chicago and London: University of Chicago Press, 2013).
Shapin, Steven, "Property, Patronage, and the Politics of Science: the Founding of the Royal Society of Edinburgh," *British Journal for the History of Science*, 7 (1974), 1-41.
Shugg, Wallace, "The Cartesian Beast-Machine in English Literature (1663-1750)," *Journal of the History of Ideas*, 29 (1968), 279-92.
—, "Humanitarian Attitudes in the Early Animal Experiments of the Royal Society," *Annals of Science*, 24 (1968), 227-38.
Singer, Peter, *Animal Liberation* (New York: Avon Books, 1975).
Sluijter, Eric J., "Didactic and Disguised Meanings? Several Seventeenth-Century Texts on Painting and the Iconological Approach to Northern Dutch Painting of this Period," in *Art in History, History in Art: Studies in Seventeenth-Century Dutch Culture*, ed. by David Freedberg and Jan de Vries (Santa Monica: Getty Center for the History of Art and the Humanities, 1991), 175-207.
Smith, Richard A., "The Eco-Suicidal Economics of Adam Smith," *Capitalism Nature Socialism*, 18 (2007), 22-43.
Sorabji, Richard, *Animal Minds and Human Morals: The Origins of the Western Debate* (Ithaca: Cornell University Press, 1993).
Sørensen, Madeleine Pinault, "Portraits of Animals, 1600-1800," in *A Cultural History of Animals in the Age of Enlightenment*, ed. by Matthew Senior (Oxford and New York: Berg, 2007), 157-96.
Spadafora, David, *The Idea of Progress in Eighteenth-Century Britain* (New Haven and London: Yale University Press, 1990).
Spary, E. C., *Utopia's Garden: French Natural History from Old Regime to Revolution* (Chicago and London: University of Chicago Press, 2000).
Spicer, Joaneath A., "'De Koe voor d'aerde statt': The Origins of the Dutch Cattle Piece," in *Essays in Northern European Art, Presented to Egbert Haverkamp-Begemann on his Sixtieth Birthday*, ed. by Anne-Marie Logan (Doornspijk: Davaco, 1983), 251-6.
Spink, J. S., *French Free-Thought from Gassendi to Voltaire* (New York: Athlone Press, 1960).
Starobinski, Jean, *Jean-Jacques Rousseau: Transparency and Obstruction*, trans. by Arthur Goldhammer (Chicago: University of Chicago Press, 1988).
Stechow, Wolfgang, "Jan Baptist Weenix," *Art Quarterly*, 11 (1948), 180-99.

Steiguer, J. Edward de, *The Origins of Modern Environmental Thought* (Tucson: University of Arizona Press, 2006).

Steinberg, Leo, "Leonardo's Last Supper," *Art Quarterly*, 36 (1973), 297-410.

Steiner, Gary, *Anthropocentrism and Its Discontents: The Moral Status of Animals in the History of Western Philosophy* (Pittsburgh: University of Pittsburgh Press, 2005).

Strickland, Lloyd, "God's Creatures? Divine Nature and the Status of Animals in the Early Modern Beast-Machine Controversy," *International Journal of Philosophy and Theology*, 74 (2013), 291-309.

Stroup, Alice, "Christiaan Huygens & the Development of the Air Pump," *Janus*, 68 (1981), 129-58.

Sullivan, Scott A., *The Dutch Gamepiece* (Montclair, NJ: Allanheld & Schram, 1984).

—, "Jan Baptist Weenix: 'Still Life with a Dead Swan'," *Bulletin of the Detroit Institute of Arts*, 57 (1979), 64-71.

—, "Rembrandt's 'Self-Portrait with a Dead Bittern'," in *Art Bulletin*, 62 (1980), 236-43.

Sutton, Elizabeth, *Art, Animals, and Experience: Relationships to Canines and the Natural World* (New York and London: Routledge, 2017).

Sutton, Peter, "The Noblest of Livestock," *J. Paul Getty Museum Journal*, 15 (1987), 97-110.

Swan, Claudia, "Ad Vivum, Naer het Leven, From the Life: Defining a Mode of Representation," in *Word & Image*, 11 (1995), 353-72.

Tague, Ingrid H., *Animal Companions: Pets and Social Change in Eighteenth-Century Britain* (University Park: Pennsylvania State University Press, 2015).

Thirsk, Joan, "Horses in Early Modern England: For Service, for Pleasure, for Power," in *The Rural Economy of England: Collected Essays* (London: Hambledon Press, 1984), 375-401.

Thomas, Janice, "Does Descartes Deny Consciousness to Animals?" *Ratio*, 19 (2006), 336-63.

Thomas, Keith, *Man and the Natural World: Changing Attitudes in England 1500-1800* (Harmondsworth: Penguin Books, 1984).

Tijn, Th. van, "Dutch Economic Thought in the Seventeenth Century," in *Economic Thought in the Netherlands: 1650-1950*, ed. by J. van Daal and A. Heertje (Aldershot: Avebury, 1992), 7-28.

Tomaselli, Sylvana, "Moral Philosophy and Population Questions in Eighteenth Century Europe," *Population and Development Review*, 14, Supplement (1988), 7-29.

Tranöy, Knut Erik, "Hume on Morals, Animals, and Men," *Journal of Philosophy*, 56 (1959), 94-103.

Tucker, Joseph, "The *Turkish Spy* and its French Background," *Revue de littérature compare*, 32 (1958), 74-91.

Turnbull, Alexandra, "The Horse in Landscape: Animals, Grooming, Labour and the City in the Seventeenth-Century Netherlands," *Shift*, 3 (2012), 1-24.

Turner, James, *Reckoning with the Beast: Animals, Pain, and Humanity in the Victorian Mind* (Baltimore and London: Johns Hopkins University Press, 1980).

Vardi, Liana, *The Physiocrats and the World of the Enlightenment* (Cambridge: Cambridge University Press, 2012).

Vergara, Lisa, *Rubens and the Poetics of Landscape* (New Haven and London: Yale University Press, 1982).

Verslype, Ige, "A Preliminary Study on Paulus Potter's (1625-1654) Painting Technique," *Art Matters*, 3 (2005), 97-110.

Walsh, Amy, *Paulus Potter: His Works and Their Meaning*, Ph.D. diss. (New York: Columbia University, 1985).

Walsh, Amy, Edwin Buijsen, and Ben Broos, *Paulus Potter: Paintings, Drawings and Etchings*, exhibition catalogue (The Hague: Mauritshuis, and Zwolle: Waanders, 1994).

Watson, Robert N., *Back to Nature: The Green and the Real in the Late Renaissance* (Philadelphia: University of Pennsylvania Press, 2006).

Weinstein, Jack Russell, "What My Dog Can Do: On the Effect of *The Wealth of Nations* I.ii.2," in *Propriety and Prosperity: New Studies on the Philosophy of Adam Smith*, ed. by David F. Hardwick and Leslie Marsh (Basingstoke: Palgrave Macmillan, 2014), 147-65.

Wells, David, "Resurrecting the Dismal Parson: Malthus, Ecology, and Political Thought," *Political Studies*, 30 (1982), 1-15.

Westman, Robert S., "Huygens and the Problem of Cartesianism," in *Studies on Christiaan Huygens: Invited Papers from the Symposium on the Life and Work of Christiaan Huygens, Amsterdam, 22-25 August 1979*, ed. by H. J. M. Bos, M. J. S. Rudwick, H. A. M. Snelders, and R. P. W. Visser (Lisse: Swets & Zeitlinger, 1980), 83-103.

Westrheene, T. van, *Paulus Potter: sa vie et ses oeuvres* (The Hague: Nijhoff, 1867).

Weststeijn, Arthur, *Commercial Republicanism in the Dutch Golden Age: The Political Thought of Johan & Pieter de la Court* (Leiden and Boston: Brill, 2012).

Wettengl, Kurt, ed., *Maria Sibylla Merian 1647-1717: Artist and Naturalist* (Ostfildern: Verlag Gerd Hatje, 1998).

Wheelock, Arthur K., Jr., *Aelbert Cuyp*, exhibition catalogue (Washington, DC: Thames & Hudson, 2001).

Wheelock, Arthur K., Jr., with contributions by Alison McNeil Kettering, Arie Wallert, and Marjorie E. Wieseman, *Gerard ter Borch*, exhibition catalogue (Washington D.C.: National Gallery of Art, and New Haven: Yale University Press, 2004).

Whelan, Frederick G., "Population and Ideology in the Enlightenment," *History of Political Thought*, 12 (1991), 35-72.

Whitehead, P. J. P., "Georg Markgraf and Brazilian Zoology," in *Johan Maurits van Nassau-Siegen: 1604-1679: A Humanist Prince in Europe and Brazil*, ed. by E. van den Boogaart, H. R. Hoetink, and P. J. P. Whitehead (The Hague: Johan Maurits van Nassau Stichting, 1979), 424-71.

Wind, Barry, "Annibale Carracci's 'Scherzo': The Christ Church 'Butcher Shop'," *Art Bulletin*, 58 (1976), 93-6.

Witztum, Amos, "Property Rights and the Right to the Fruits of One's Labor: A Note on Adam Smith's Jurisprudence," *Economics and Philosophy*, 21 (2005), 279-89.

Wolloch, Nathaniel, "Animals, Extraterrestrial Life and Anthropocentrism in the Seventeenth Century," *The Seventeenth Century*, 17 (2002), 235-53.

—, *Animals in the Netherlands: Human Attitudes toward Animals as Reflected in the Culture and Art of the Seventeenth-Century Netherlands* (Ph.D. diss., The Hebrew University of Jerusalem, 1997).

—, "Edward Gibbon's Cosmology," *International Journal of the Classical Tradition*, 17 (2010), 165-77.

—, "'Facts, or Conjectures': Antoine-Yves Goguet's Historiography," *Journal of the History of Ideas*, 68 (2007), 429-49.

—, *History and Nature in the Enlightenment: Praise of the Mastery of Nature in Eighteenth-Century Historical Literature* (Farnham and Burlington: Ashgate, 2011).

—, "History 'On a New Plan': Robert Henry (1718-1790) and the Rise of Interdisciplinary Historiography," *Storia della Storiografia*, 58 (2010), 35-52.

—, "Iconography, Style and the Historical Study of Art," *Storia della Storiografia*, 45 (2004), 109-24.

—, "The Liberal Origins of the Modern View of Nature," *Tocqueville Review*, 34 (2013), 107-31.

—, "Natural Disasters and the Debate on the Unity or Plurality of Enlightenments," *The Eighteenth Century: Theory and Interpretation*, 57 (2016), 325-42.

—, *Nature in the History of Economic Thought: How Natural Resources Became an Economic Concept* (London and New York: Routledge, 2017).

—, *Subjugated Animals: Animals and Anthropocentrism in Early Modern European Culture* (Amherst, NY: Humanity Books, 2006).

Womack, William R., "Eighteenth-Century Themes in the *Histoire philosophique et politique des deux Indes* of Guillaume Raynal," *Studies on Voltaire and the Eighteenth Century*, 96 (1972), 129-265.

Womersley, David, *The Transformation of the Decline and Fall of the Roman Empire* (Cambridge: Cambridge University Press, 1988).

Wood, P. B., "Buffon's Reception in Scotland: The Aberdeen Connection," *Annals of Science*, 44 (1987), 169-90.

—, "The Natural History of Man in the Scottish Enlightenment," *History of Science*, 28 (1990), 89-123.

Wood, Paul, "Introduction," in William Smellie, *The Philosophy of Natural History*, 2 vols. (Edinburgh and London: various publishers, 1790, 1799; repr. Bristol and Sterling: Thoemmes Press, 2001).

Wootton, David, "David Hume, 'the historian'," in *The Cambridge Companion to Hume*, ed. by David Fate Norton (Cambridge: Cambridge University Press, 1993), 281-312.

Worster, Donald, *Nature's Economy: A History of Ecological Ideas*, 2nd ed. (Cambridge: Cambridge University Press, 1994).

Wyett, Jodi L., "The Lap of Luxury: Lapdogs, Literature, and Social Meaning in the 'Long' Eighteenth Century," *LIT: Literature Interpretation Theory*, 10 (1999), 275-301.

Young, Jeffrey T., "Natural Jurisprudence and the Theory of Value in Adam Smith," *History of Political Economy*, 27 (1995), 755-73.

Zumthor, Paul, *Daily Life in Rembrandt's Holland*, trans. by Simon W. Taylor (London: Weidenfeld, 1962).

Index